WEF/ABC/C_2EP

Collection Systems Operators'

Guide to Preparing for the Certification Examination

2014
Water Environment Federation
601 Wythe Street
Alexandria, VA 22314-1994
www.wef.org

ISBN 978-1-57278-301-0

IMPORTANT NOTICE

WATER ENVIRONMENT FEDERATION

Founded in 1928, the Water Environment Federation (WEF) is a not-for-profit technical and educational organization with members from varied disciplines who work toward the WEF vision of preservation and enhancement of the global water environment. The WEF network includes more than 100,000 water quality professionals from 77 Member Associations in 31 countries.

For information on membership, publications, and conferences, contact

Water Environment Federation
601 Wythe Street
Alexandria, VA 22314-1994 USA
1-703-684-2400
http://www.wef.org

ASSOCIATION OF BOARDS OF CERTIFICATION

Established in 1972, the Association of Boards of Certification (ABC) is a non-profit member-driven organization dedicated to protecting public health and the environment by advancing the quality and integrity of environmental certification programs. Today, ABC's membership includes almost 100 certifying authorities, representing more than 40 states, 10 Canadian provinces and territories, and several international programs. More than 96 certification programs currently use ABC examinations to test more than 30,000 candidates annually.

The ABC vision is as follows: Promote integrity in environmental certification throughout the world.

The ABC mission is as follows: The Association of Boards of Certification is dedicated to advancing the quality and integrity of environmental certification programs.

Association of Boards of Certification objectives:

- Promote certification as a means of protecting public health, the infrastructure, and the environment;
- Promote uniformity of standards and best practices in certification;
- Serve as a technical resource for certification entities;
- Facilitate the transfer of certification between certifying authorities; and
- Serve the needs of our members.

CERTIFICATION COMMISSION FOR ENVIRONMENTAL PROFESSIONALS

In 2012, ABC created the Certification Commission for Environmental Professionals (C_2EP) to enhance and administer ABC's certification and standardized exam development programs. The Certification Commission for Environmental Professionals is composed of industry stakeholders and independently administers professional operator (PO) certifications and professional designations built on the highest standards for psychometrics and personnel certification. The C_2EP mission is as follows: C_2EP is committed to protecting public health and the environment through certification and professional designation of water industry operators.

The Association of Boards of Certification and C_2EP have identified the following WEF manuals as primary reference material to help operators prepare for the ABC/C_2EP collection system operator certification exams:

- *Operation of Municipal Wastewater Treatment Plants*—Manual of Practice No. 11, 6th Edition
- *Existing Sewer Evaluation and Rehabilitation*—Manual of Practice No. FD-6, 3rd Edition
- *Wastewater Collection Systems Management*—Manual of Practice No. 7, 6th Edition

In addition to the primary references listed above, WEF has many other technical resources that are listed in the study guide. If you are having difficulty answering any of the questions in the study guide, you should consult the reference source provided following the answer to the question.

For information on PO certification and professional designation, testing services, or other ABC/C_2EP opportunities, contact

Association of Boards of Certification/Certification Commission for Environmental Professionals
2805 SW Snyder Blvd, Suite 535
Ankeny, IA 50023 USA
1-515-232-3623
http://www.abccert.org
abc@abccert.org

CONTENTS

PREFACE

The Association of Boards of Certification (ABC), the Certification Commission for Environmental Professionals (C_2EP), and the Water Environment Federation (WEF) have worked together to produce this document based on industry-wide job analyses and *Need-to-Know Criteria.* This publication is designed to give the user practice answering multiple choice examination questions that will be similar in format and content to those they might find on the certification exam.

The questions test skills and knowledge required of an operator working in a wastewater collection system. Some of the questions are more difficult and complex than others.

WEF/ABC/C_2EP Disclaimer

The questions included in this guide have been chosen to sample as many different aspects of a collection system operator's job responsibilities as possible. However, because of the tremendous variety in equipment, processes, conditions, and operator duties, not all of the questions may be useful in all of the possible certification applications. It should also be noted that questions in the study guide are only intended to provide an example of style and possible topics for certification exam questions and will not be found on an ABC/C_2EP certification exam.

The following people participated in the development of *WEF/ABC/C_2EP Collection System Operators' Guide to Preparing for the Certification Examination*:

Task Force Chair, Bradley E. Fix, Superintendent, City of Shelbyville, Shelbyville, Indiana

Authors:

Samantha E. Bartow, Director of Sewer Services, Taylors Fire and Sewer District, Taylors, South Carolina
Laurie Chase, President, Blue Heron Engineering Services, Ltd., Dublin, Ohio
Steve Donovan, Wastewater Collections Systems O&M Specialist, Brown and Caldwell, Columbus, Ohio
Brent Leigh Erickson, Wastewater Operations Manager, City of Garland, Garland, Texas
Scott M. Firmin, Director of Wastewater Services, Portland Water District, Portland, Maine
David A. Flowers, Natural Water Solutions LLC, Cedarburg, Wisconsin
Michael Harmer, Development Services Group Leader, Washington Suburban Sanitary Commission, Laurel, Maryland
Louis Lambe, Superintendent of Line Operations, Passaic Valley Sewerage Commissioners, Newark, New Jersey
Robert John Potter, Wastewater Manager II, City of Los Angeles, Los Angeles, California
John T. Shaw, John Shaw Consulting, LLC

Stephen Tilson, Tilson and Associates LLC, Torrington, Connecticut
Robert A. Villée, Executive Director, Plainfield Area Regional Sewerage Authority, Middlesex, New Jersey
Michel Wanna, Field Operations Director, Macon Water Authority, Macon, Georgia

Reviewers:

Mike Bell, Water Distribution Class IV, Water Treatment Class III, Wastewater Treatment Class III, Wastewater Collection Class III (Writing Class IV in November), Corporation for the City of Hamilton, Hamilton, Ontario, Canada
Darryl Macy, Macon Water Authority, Macon, Georgia
Stacy J. Passaro, P.E., BCEE, Passaro Engineering LLC, Mount Airy, Maryland
Lonn Rasmussen, Grade 4 Operator, Cottonwood Improvement District, Sandy, Utah
Coby Shurtleff, MBA, CWP, Operations Manager, Town of Windsor, Public Works, Windsor, Colorado
H. William Steinhauser, Jr., State of Georgia Wastewater Collection License, State of Georgia Wastewater Treatment License Class III, City of Savannah, Savannah, Georgia
Bill Weitkemper, Sewer Maintenance Superintendent (Retired), City of Columbia, Columbia, Missouri

CERTIFICATION COMMISSION FOR ENVIRONMENTAL PROFESSIONALS EDUCATION AND EXPERIENCE REQUIREMENTS

The Certification Commission for Environmental Professionals offers certification and professional designation to collection system operators. In addition to examination requirements, the following education and experience requirements must be met for C_2EP certification and professional designation:

CLASS I

- High school diploma, general equivalency diploma (GED), or equivalent;
- One year of acceptable operating experience; and
- Ninety contact hours of postsecondary education.

CLASS II

- High school diploma, GED, or equivalent;
- Three years of acceptable experience in a Class I or higher utility; and
- One hundred eighty contact hours of postsecondary education.

CLASS III

- High school diploma, GED, or equivalent;
- Four years of acceptable operating experience in a Class II or higher utility, including two years of direct responsible charge; and
- Nine hundred contact hours of postsecondary education.

CLASS IV

- High school diploma, GED, or equivalent;
- Four years of operating experience in a Class III or higher utility, including two years of direct responsible charge; and
- One thousand eight hundred contact hours of postsecondary education.

In addition, ABC and C_2EP also provide testing and certifications services to many states and provinces. In these instances, the individual state or province may determine the eligibility requirements and/or exam content. Contact your certification authority directly to receive the latest information regarding their specific eligibility requirements and exam specifications.

IMPORTANCE OF CERTIFICATION

The wastewater collection system operator is responsible for taking actions to protect public health and the environment by ensuring that our wastewater collection system infrastructure is operating properly and that this infrastructure is maintained and protected to achieve a maximum usable life.

Today's collection system operators must understand how each component of a system functions and how these components come together in a hydraulically connected system to collect and transport wastewater to downstream treatment facilities. Collection system operators must also understand the effects that the operation and performance of a collection system have on the function and performance of increasingly more complex treatment facilities with more stringent discharge permits. Properly operating and maintaining a collection system requires knowledge and competency in a broad range of topics and skills. Certification provides operators with a third-party verified yardstick of achievement that demonstrates their level of accumulated knowledge, skills, and experience that are important in the day-to-day activities their jobs require.

Many utility owners and managers place value and importance on an operator's steady progress through the various levels of certification. It demonstrates a focus and commitment to building knowledge and skills that improve decision making and actions on the job.

The following passage was written by a lifelong collection system operator when he was asked about the importance of certification from an operator's perspective:

WHY SHOULD YOU BECOME A CERTIFIED WASTEWATER COLLECTION SYSTEM OPERATOR?

> I have been a Wastewater Collection System Operator since 1976, 2 years after I began employment with a large regional wastewater agency in the Minneapolis/St. Paul MN metropolitan area. Minnesota, at the time, was one of the few states that required Collection System Certification. I worked in the Regional Collection System and when I found out about not only the Certification requirements, but also the state sponsored 3 day Annual Collection System Certification Training program, I could see the benefits to my career.
>
> First, by passing the highest grade of certification, the SA, my employer knew that I had the prerequisite professional knowledge to operate, maintain and manage a large wastewater Collection System. Second, I fulfilled the State Regulatory Requirement for a Certified Operator. Third, I was encouraged to become involved in Certification Training for fellow agency employees. Certification would, in the future, be a requirement for employment at this agency. Certification at 1 of the 4 levels also resulted in a certified employee receiving "Certification Pay" that increased as the certification level increased. This not only added [to] their paycheck every 2 weeks, but also added to their long-term retirement benefits.
>
> I felt so strongly about Collection System Operator Certification that I became involved with local, state and national professional organizations, for

example, the Water Environment Federation and Member Associations in Minnesota, Wisconsin and California. It has benefitted me substantially in many ways, throughout my almost 40 year career in Wastewater Collection Systems. Even though I am semi-retired, I still maintain my Certification, which I am very proud of.

I believe that Wastewater Collection System Certification is even more important today as most states now require some level of voluntary or mandatory Operator Certification. Also, many more employers now require Certified Operators as a condition of employment.

In summary, Certification will benefit you professionally and financially.

Rick Arbour
Minnesota Pollution Control Agency
Wastewater Collection System Operator Certificate SA-4061

TIPS FOR TAKING A CERTIFICATION EXAMINATION

The following are some helpful tips for successfully taking a certification examination.

BE PREPARED

The first step in passing a certification examination is to prepare for the test in advance by studying the type of information that will be covered in the examination. One method of preparing for the test is to attend training classes offered by local wastewater utilities, community colleges, and vocational-technical schools. Another useful method is to read and study reference books on wastewater treatment. Many state certification agencies publish or can recommend reference materials that can help you study the material covered on the examination in your state. In addition, some state or provincial environmental departments provide sample tests, lists of material covered in the examinations, or examples of typical test problems. Each question in this guide is referenced to a technical source. If you find you are having difficulty with a specific topic, you may want to review the material in that reference.

If you are preparing for the exam on your own, it is critical that you work out a doable schedule that will allow you ample time to work through each of the important topic areas that the exam will cover. To do this, review each of the core competencies and decide what material you want to review for each and how many sample questions you will take for each competency. Determine how many hours of study time will be required to complete the review and practice questions. On your schedule, working back from the date when you would like to take the exam, schedule blocks of study time on specific days for each topic for the total amount of time required. Make sure you stick with this schedule as you prepare for the exam. Do not plan to "cram" or try to study all the material in the week or so before the exam. This is not an effective way to study.

Certification examinations are intended to test every aspect of an operator's involvement in collection system operation and maintenance. Because calculations are often involved, certification examinations include mathematical problems dealing with flowrates, pump performance, wet well volumes, and so on. To prepare for the mathematical problems, it is a good idea to memorize the more common conversion factors (constants) used in collection systems mathematics. Formula/conversion tables are handed out at ABC/C_2EP exams. A version of this table is included at the end of this section.

Remember that examinations are changed frequently and computers can be used to generate random exams from larger sets of questions. Therefore, do not try to learn the answers to specific problems that you may have heard about from other operators. Rather, concentrate on learning how to work various types of problems, such as how to calculate a target return sludge rate or how to adjust a chemical feed pump. In this way, you will be prepared to handle any question on the test, not just certain questions you have memorized.

Although operators are responsible for knowing all aspects of wastewater treatment, certification tests in different regions emphasize some aspects more than others. This guide is broken into topics that are commonly emphasized in the examinations. By talking with your certification program coordinator, you can learn which additional areas may be emphasized in the examination given in your area.

As a final tip for preparing for the test, give yourself the night before the exam off from studying. Do something relaxing and try to get to bed early and get a good night's rest. Do not try to "cram" information into your brain during the last few hours. Cramming typically is not beneficial and can lead to confusion and exhaustion.

TAKING THE EXAMINATION

On the day of the examination, be at the testing location well ahead of the announced starting time. Be sure that you have the necessary materials, such as a spare pencil and your admittance slip (if required). Also, remember to put fresh batteries in your calculator because working mathematical calculations by hand can waste valuable time. Before starting the test, skim through the questions to get a general idea of the kinds of questions that are on the test. By seeing the types of questions and math problems you must answer at the start, you will have a better idea about how to use your time. It is a good idea to answer all the easy questions first and then tackle the more complex problems that will take longer to complete.

If the test has essay questions and lengthy math problems, think about the person who will grade these questions. To help the grader, write your answer clearly and in an order that the grader can easily follow. Remember that graders are human and that they will be reading lots of essays and math problems. Although they may not intentionally lower your grade for sloppy or unreadable work, such work certainly will not help your score. Underline answers to math problems—do not make the grader search through pages of figures to find your answer.

CALCULATION PROBLEMS

Because calculation problems are an important and often difficult part of certification examinations, refer to the following guidelines when solving calculation problems:

- Read each question carefully to ensure that you know what answer is required;
- Make a drawing or sketch if it will help to solve the problem;
- Simplify the problem. If the problem is complex, break it down into small pieces that you can solve separately;
- Make all necessary conversions (such as converting 1.5 ft to 18 in., if the question requires an answer in inches);
- Give your answer only to the nearest significant figure. For example, a chemical dosage problem with an answer of 26.93276 lb is meaningless when the chemical feeder is only accurate to the nearest pound;
- Be sure the decimal point is in the right place; and
- Check to see that your answer makes sense. For example, if you get an answer of $30,000,000,000 for a question that asks for the annual cost of odor control chemicals for a 0.5-mgd pumping station, your answer is obviously wrong. Try to find another approach to solving this math problem.

HELPFUL HINTS

Most certification examinations are mainly composed of multiple-choice questions. You should never leave a multiple-choice question blank. Often, you can determine that one or two of the potential answers are obviously wrong. If you narrow your choice to only two possible answers, you have a 50/50 chance of getting the right answer, no matter which answer you choose.

Take the examination methodically and deliberately. Avoid the natural temptation to rush through it. Be relaxed while taking the examination. Periodically, while sitting in your seat, take a moment to stretch your neck and shoulders to ease the tension.

EXAMINATION CHECKLIST

Before leaving your home or work to go to the testing location, be sure that you have the following items so that you will be well prepared to take the examination:

- An admittance slip (if one is required),
- Two sharpened no. 2 pencils with good erasers,
- A calculator,
- Fresh batteries for the calculator,
- A watch,
- Glasses (if you wear them), and
- Confidence.

Remember to stay calm and to carefully work your way through the questions. Trust your instincts and, if you have properly prepared for the examination, you should do well.

Best of luck with your examination and your career as a collection system operations professional!

Formula/Conversion Table for Wastewater Treatment, Industrial, Collection, and Laboratory Exams

*Pie wheel format for certain equations available at the end of this section.

$$\text{Alkalinity, as mg } CaCO_3\text{/L} = \frac{\text{(Titrant volume, mL)(Acid normality)(50 000)}}{\text{Sample volume, mL}}$$

$$\text{Amps} = \frac{\text{Volts}}{\text{Ohms}}$$

$$\text{*Area of circle} = (0.785)(\text{Diameter}^2)$$
$$= (\pi)(\text{Radius}^2)$$

$$\text{Area of cone (Lateral area)} = (\pi)(\text{Radius})\sqrt{\text{Radius}^2 + \text{Height}^2}$$

$$\text{Area of cone (Total surface area)} = (\pi)(\text{Radius})\left(\text{Radius} + \sqrt{\text{Radius}^2 + \text{Height}^2}\right)$$

$$\text{Area of cylinder (Total exterior surface area)} = [\text{Surface area of end \#1}] + [\text{Surface area of end \#2}] + [(\pi)(\text{Diameter})(\text{Height or depth})]$$

$$\text{*Area of rectangle} = (\text{Length})(\text{Width})$$

$$\text{*Area of right triangle} = \frac{(\text{Base})(\text{Height})}{2}$$

$$\text{Average (arithmetic mean)} = \frac{\text{Sum of all terms}}{\text{Number of terms}}$$

Average (geometric mean) $= [(X_1)(X_2)(X_3)(X_4)(X_n)]^{1/n}$ The *n*th root of the product of *n* numbers

$$\text{Biochemical oxygen demand (Unseeded), mg/L} = \frac{[\text{(Initial dissolved oxygen, mg/L)} - \text{(Final dissolved oxygen, mg/L)}][300\text{ mL}]}{\text{Sample volume, mL}}$$

$$\text{Chemical feed pump setting, \% stroke} = \frac{\text{Desired flow}}{\text{Maximum flow}} \times 100\%$$

$$\text{Chemical feed pump setting, mL/min} = \frac{\text{(Flow, mgd)(Dose, mg/L)(3.785 L/gal)(1 000 000 gal/mil. gal)}}{\text{(Liquid, mg/mL)(24 hr/d)(60 min/hr)}}$$

$$\text{Circumference of circle} = (\pi)(\text{Diameter})$$
$$= 2(\pi)(\text{Radius})$$

$$\text{Composite sample single portion} = \frac{(\text{Instantaneous flow})(\text{Total sample volume})}{(\text{Number of portions})(\text{Average flow})}$$

$$\text{Cycle time, min} = \frac{(\text{Storage volume, gal})}{(\text{Pump capacity, gpm} - \text{Wet well inflow, gpm})}$$

$$\text{Degrees Celsius} = (\text{Degrees Fahrenheit} - 32)(\tfrac{5}{9}) = \frac{(°\text{F} - 32)}{1.8}$$

$$\text{Degrees Fahrenheit} = (\text{Degrees Celsius})(\tfrac{9}{5}) + 32 = (\text{Degrees Celsius})(1.8) + 32$$

$$\text{Detention time} = \frac{\text{Volume}}{\text{Flow}}$$ *Units must be compatible*

$$\text{Dose} = \text{Demand} + \text{Residual}$$

*Electromotive force (EMF), volts = (Current, amps)(Resistance, ohms) *or* $E = IR$

$$\text{*Feed rate, lb/d} = \frac{(\text{Dosage, mg/L})(\text{Capacity, mgd})(8.34\ \text{lb/gal})}{\text{Purity, \% expressed as a decimal}}$$

$$\text{Filter backwash rise rate, in./min} = \frac{(\text{Backwash rate, gpm/sq ft})(12\ \text{in./ft})}{7.48\ \text{gal/cu ft}}$$

$$\text{Filter flowrate or backwash rate, gpm/sq ft} = \frac{\text{Flow, gpm}}{\text{Filter area, sq ft}}$$

$$\text{Filter yield, lb/hr/sq ft} = \frac{(\text{Solids loading, lb/d})(\text{Recovery, \% expressed as a decimal})}{(\text{Filter operation, hr/d})(\text{Area, sq ft})}$$

*Flowrate, cfs = (Area, sq ft)(Velocity, ft/sec) *or* $Q = AV$ *Units must be compatible*

$$\text{Food-to-microorganism ratio} = \frac{\text{biochemical oxygen demand (BOD)}_5\text{, lb/d}}{\text{mixed liquor volatile suspended solids (MLVSS), lb}}$$

*Force, lb = (Pressure, psi)(Area, sq in.)

$$\text{Gallons/capita/day} = \frac{\text{Volume of water produced, gpd}}{\text{Population}}$$

$$\text{Hardness, as mg CaCO}_3\text{/L} = \frac{\text{(Titrant volume, mL)(1000)}}{\text{Sample volume, mL}}$$ *Only when the titration factor is 1.00 of EDTA*

$$\text{Horsepower, brake (bhp)} = \frac{\text{(Flow, gpm)(Head, ft)}}{\text{(3960)(Pump efficiency, \% expressed as a decimal)}}$$

$$\text{Horsepower, motor (mhp)} = \frac{\text{(Flow, gpm)(Head, ft)}}{\text{(3960)(Pump efficiency, \% expressed as a decimal)(Motor efficiency, \% expressed as a decimal)}}$$

$$\text{*Horsepower, water (whp)} = \frac{\text{(Flow, gpm)(Head, ft)}}{3960}$$

$$\text{Hydraulic loading rate, gpd/sq ft} = \frac{\text{Total flow applied, gpd}}{\text{Area, sq ft}}$$

$$\text{Leakage, gpd} = \frac{\text{Volume, gal}}{\text{Time, days}}$$

*Mass, lb = (Volume, mil. gal)(Concentration, mg/L)(8.34 lb/gal)

*Mass flux, lb/d = (Flow, mgd)(Concentration, mg/L)(8.34 lb/gal)

$$\text{Mean cell residence time (MCRT) or solids retention time (SRT), days} = \frac{\text{Aeration tank total suspended solids (TSS), lb}}{\text{+ Clarifier TSS, lb TSS wasted, lb/d + Effluent TSS, lb/d}}$$

Milliequivalent = (mL)(Normality)

$$\text{Molarity} = \frac{\text{Moles of solute}}{\text{Liters of solution}}$$

$$\text{Motor efficiency, \%} = \frac{\text{Brake hp}}{\text{Motor hp}} \times 100\%$$

$$\text{Normality} = \frac{\text{Number of equivalent weights of solute}}{\text{Liters of solution}}$$

$$\text{Number of equivalent weights} = \frac{\text{Total weight}}{\text{Equivalent weight}}$$

$$\text{Number of moles} = \frac{\text{Total weight}}{\text{Molecular weight}}$$

$$\text{Organic loading rate, lb BOD}_5\text{/d/cu ft} = \frac{\text{Organic load, lb BOD}_5\text{/d}}{\text{Volume, cu ft}}$$

$$\begin{array}{c}\text{Organic loading rate-rotating biological contractor,}\\ \text{lb BOD}_5\text{/d/1000 sq ft}\end{array} = \frac{\text{Organic load, lb BOD}_5\text{/d}}{\text{Surface area of media, 1000 sq ft}}$$

$$\text{Organic loading rate-tricking filter, lb BOD}_5\text{/d/1000 sq ft} = \frac{\text{Organic load, lb BOD}_5\text{/d}}{\text{Volume, 1000 cu ft}}$$

$$\text{Oxygen uptake rate or oxygen consumption rate, mg/L/min} = \frac{\text{Oxygen use, mg/L}}{\text{Time, min}}$$

$$\text{Population equivalent, organic} = \frac{\text{(Flow, mgd)(BOD, mg/L)(8.34 lb/gal)}}{\text{BOD/d/person lb}}$$

$$\text{Recirculation ratio-tricking filter} = \frac{\text{Recirculated flow}}{\text{Primary effluent flow}}$$

$$\text{Reduction in flow, \%} = \left(\frac{\text{Original flow} - \text{reduced flow}}{\text{Original flow}}\right) \times 100\%$$

$$\text{Reduction of volatile solids, \%} = \left(\frac{\text{In} - \text{Out}}{\text{In} - (\text{In} \times \text{Out})}\right) \times 100\%$$ *All information (In and Out) must be in decimal form*

$$\text{Removal, \%} = \left(\frac{\text{In} - \text{Out}}{\text{In}}\right) \times 100\%$$

$$\text{Return rate, \%} = \frac{\text{Return flowrate}}{\text{Influent flowrate}} \times 100\%$$

$$\text{Return sludge rate-solids balance} = \frac{\text{(MLSS)(Flowrate)}}{\text{Return activated sludge suspended solids} - \text{MLSS}}$$

$$\text{Slope, \%} = \frac{\text{Drop or rise}}{\text{Distance}} \times 100\%$$

$$\text{Sludge density index} = \frac{\text{100 (Settled sludge volume}_{30}\text{, mL/L)(1000 mg/g)}}{\text{Sludge volume index}}$$

$$\text{Sludge volume index, mL/g} = \text{MLSS, mg/L}$$

$$\text{Solids, mg/L} = \frac{\text{(Dry solids, grams)(1 000 000)}}{\text{Sample volume, mL}}$$

$$\text{Solids concentration, mg/L} = \frac{\text{Weight, mg}}{\text{Volume, L}}$$

$$\text{Solids loading rate, lb/d/sq ft} = \frac{\text{Solids applied, lb/d}}{\text{Surface area, sq ft}}$$

Solids retention time: *see* MCRT

$$\text{Specific gravity} = \frac{\text{Specific weight of substance, lb/gal}}{\text{Specific weight of water, lb/gal}}$$

$$\text{Specific oxygen uptake rate (OUR) or respiration rate, (mg/g)/hr} = \frac{\text{OUR, mg/L/min (60 min)}}{\text{MLVSS, g/L (1 hr)}}$$

$$\text{Surface loading rate or surface overflow rate, gpd/sq ft} = \frac{\text{Flow, gpd}}{\text{Area, sq ft}}$$

$$\text{Three normal equation} = (N_1 \times V_1) + (N_2 \times V_2) = (N_3 \times V_3) \qquad \text{where } V_1 + V_2 = V_3$$

$$\text{Two normal equation} = N_1 \times V_1 = N_2 \times V_2 \qquad \text{where } N = \text{normality, } V = \text{volume or flow}$$

$$\text{Velocity, ft/sec} = \frac{\text{Flowrate, cfs}}{\text{Area, sq ft}} \quad \text{or} \quad \frac{\text{Distance, ft}}{\text{Time, sec}}$$

$$\text{Volatile solids, \%} = \frac{\text{Dry solids, g} - \text{Fixed solids, g}}{\text{Dry solids, g}} \times 100\%$$

$$\text{*Volume of cone} = (1/3)(0.785)(\text{Diameter}^2)(\text{Height})$$
$$= (1/3)[(\pi)(\text{Radius}^2)(\text{Height})]$$

$$\text{*Volume of cylinder} = (0.785)(\text{Diameter}^2)(\text{Height})$$
$$= (\pi)(\text{Radius}^2)(\text{Height})$$

$$\text{*Volume of rectangular tank} = (\text{Length})(\text{Width})(\text{Height})$$

$$\text{Watts (alternating current circuit)} = (\text{Volts})(\text{Amps})(\text{Power factor})$$

$$\text{Watts (direct current circuit)} = (\text{Volts})(\text{Amps})$$

$$\text{Weir overflow rate, gpd/ft} = \frac{\text{Flow, gpd}}{\text{Weir length, ft}}$$

$$\text{Weir-to-water efficiency, \%} = \frac{\text{Water horsepower, hp}}{\text{Power input, hp or motor hp}} \times 100\%$$

$$\text{Weir-to-water efficiency, \%} = \frac{(\text{Flow, gpm})(\text{Total dynamic head, ft})(0.746 \text{ kW/hp})}{(3960)(\text{Electrical demand, kW})} \times 100\%$$

Conversion Factors		
1 ac	=	43 560 sq ft
1 ac ft	=	326 000 gal
1 cfs	=	0.028 m^3/s
1 cu ft	=	7.48 gal
	=	62.4 lb
1 cfs	=	0.646 mgd
1 ft	=	0.305 m
1 ft H_2O	=	0.433 psi
1 ft/sec	=	0.305 m/s
1 gal	=	3.79 L
	=	8.34 lb
1 gpd	=	0.004 m^3/d
1 gpd/ft	=	0.012 $m^3/m{\cdot}d$
1 gpd/sq ft	=	0.041 $m^3/m^2{\cdot}d$ or m/d
1 gpm	=	0.063 L/s
1 gpm/sq ft	=	0.679 $L/m^2{\cdot}s$
	=	58.67 $m^3/m^2{\cdot}d$
1 gr/gal	=	17.1 mg/L
1 hp	=	0.746 kW
	=	746 W
	=	33 000 ft lb/min
1 in.	=	25.4 mm
1 lb/d	=	0.454 kg/d
1 lb/d/cu ft	=	16.02 $kg/m^3{\cdot}d$
1 lb/d/cu ft	=	40.74 $L/m^2{\cdot}d$
1 lb/d/sq ft	=	4.882 $kg/m^2{\cdot}d$
1 lb/gal	=	0.120 kg/L
1 lb/hr/sq ft	=	4.882 $kg/m^2{\cdot}h$
1 mile	=	5280 ft
1 mgd	=	694 gpm
	=	1.55 cfs
	=	3.785 ML/d
1 lb	=	0.454 kg
1 mil. gal	=	3.785 ML/d
1 psi	=	2.31 ft of water
	=	6.895 kPa
1 sq ft	=	0.093 m^2
1 sq in.	=	645.2 mm^2
1 ton	=	2000 lb
1%	=	10 000 mg/L
π or pi	=	3.14159

***Pie Wheels:**

- To find the quantity above the horizontal line: multiply the pie wedges below the line together.
- To solve for one of the pie wedges below the horizontal line: cover that pie wedge, then divide the remaining pie wedge(s) into the quantity above the horizontal line.

Given units must match the units shown in the pie wheel.

Area of Circle

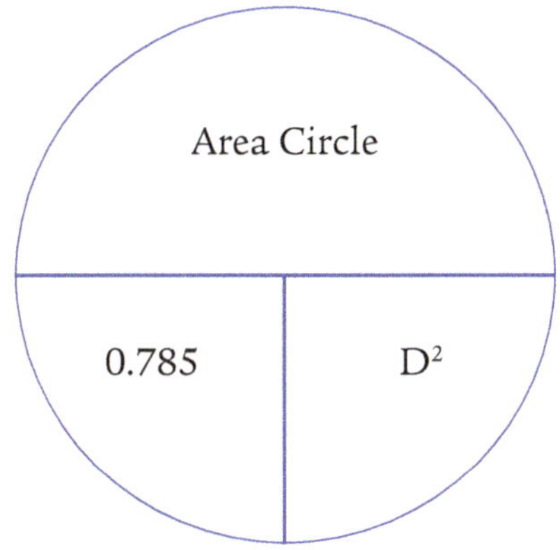

Area of Rectangle

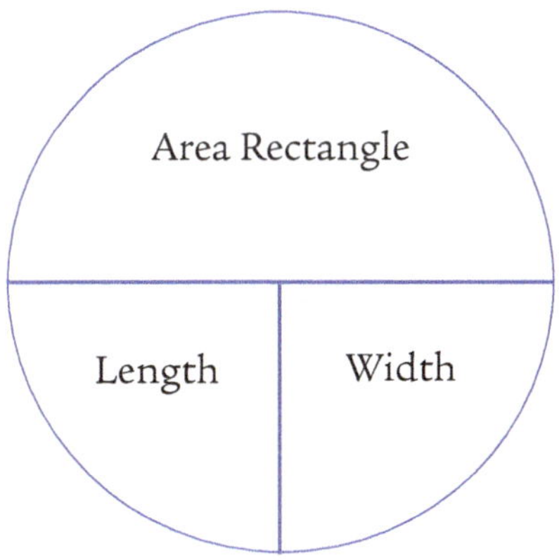

Area of Right Triangle

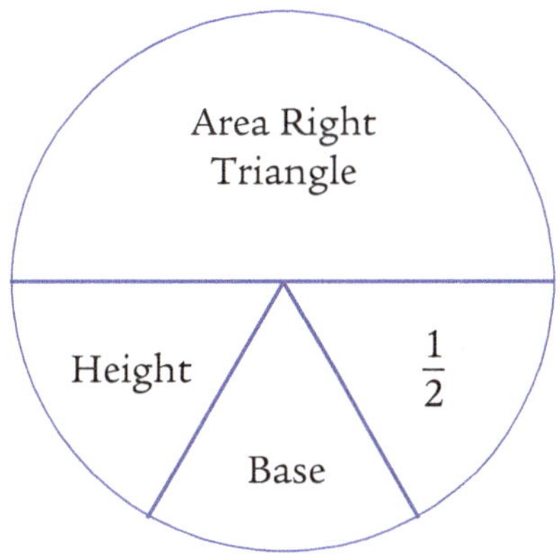

Electromotive Force (EMF), volts

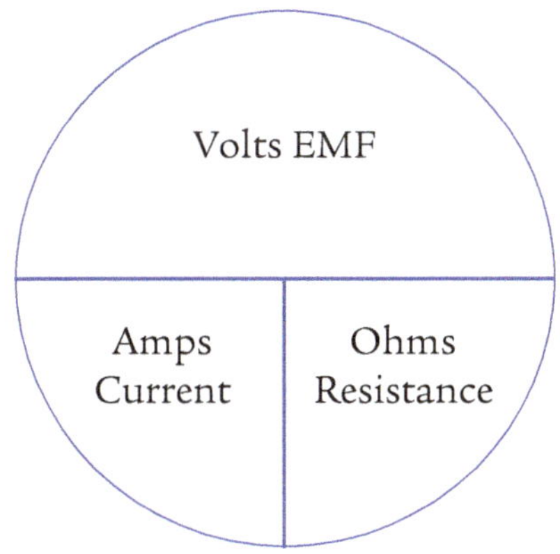

Feed Rate, lb/d

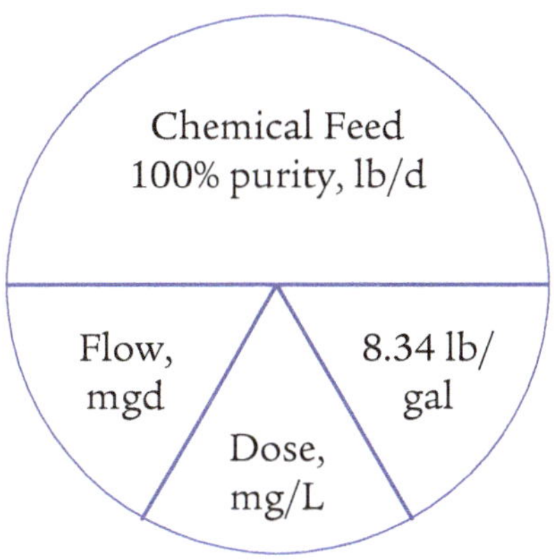

Flowrate, cfs

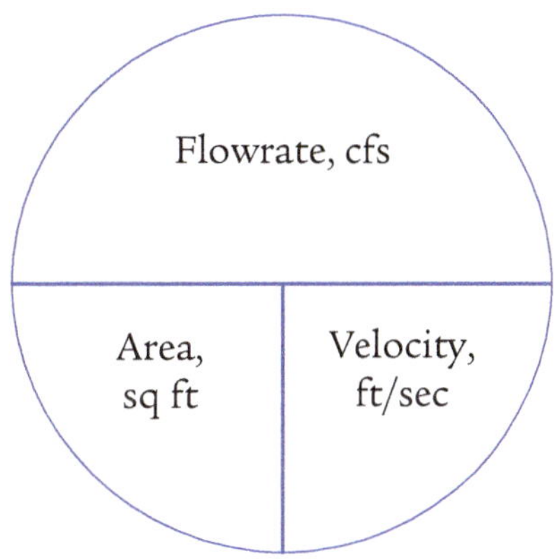

Force, lb

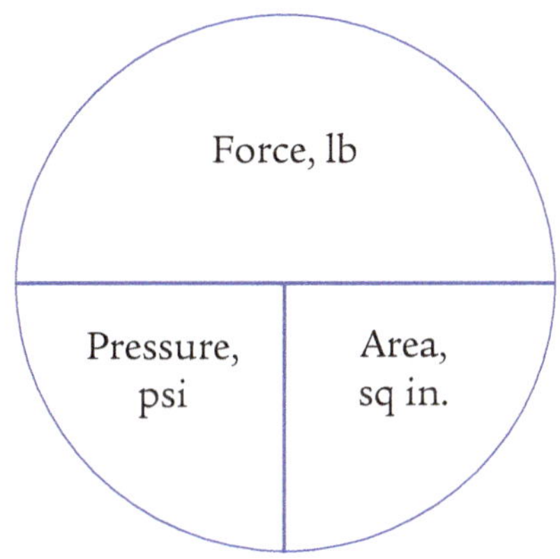

Horsepower, water

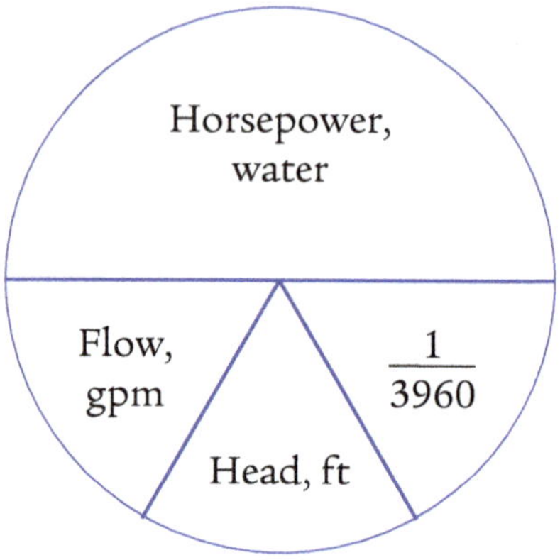

Mass, lb

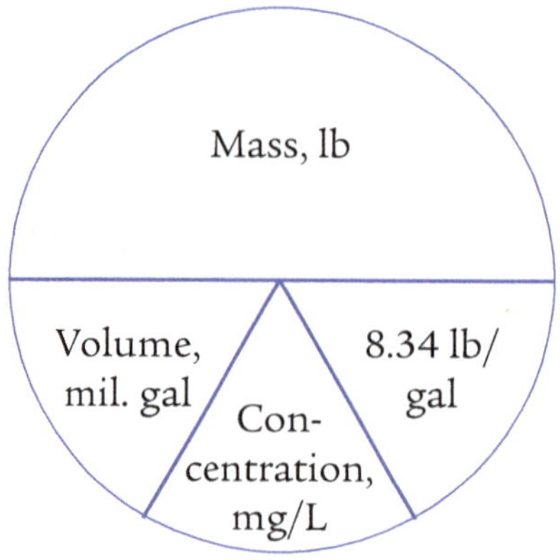

Mass Flux, lb/d

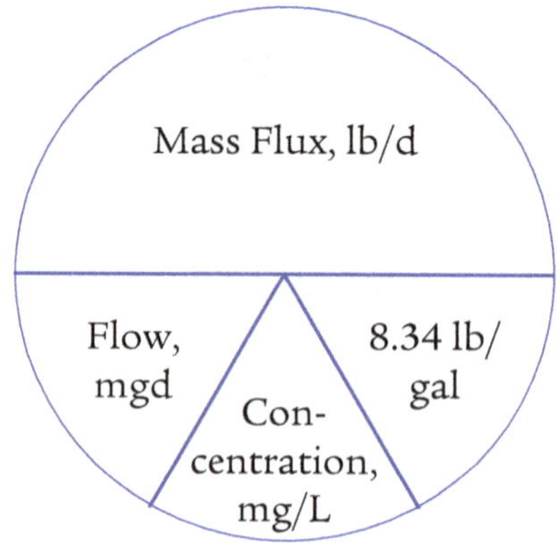

Volume of Cone

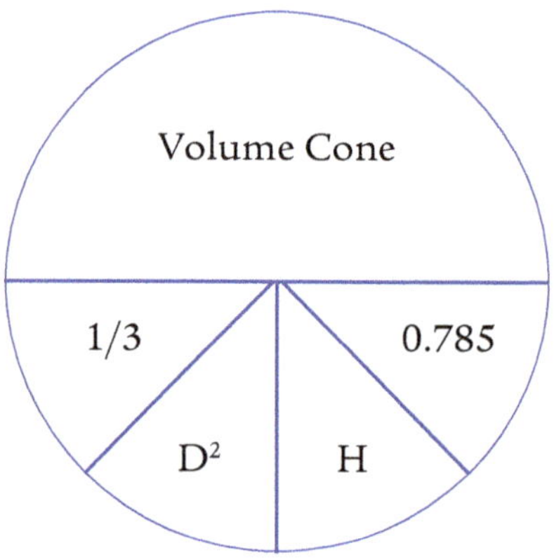

Volume of Cylinder

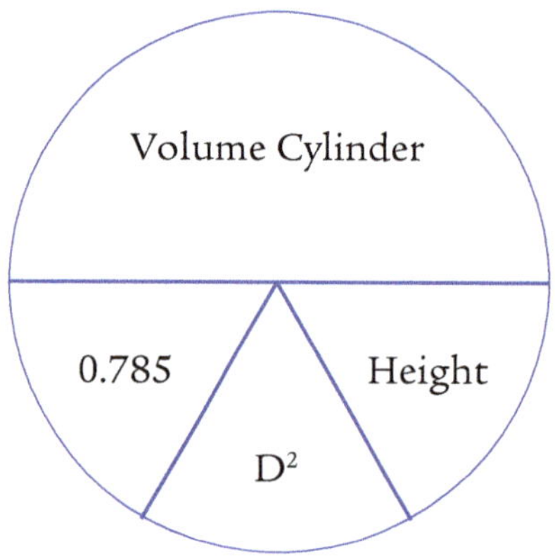

Volume of Rectangular Tank

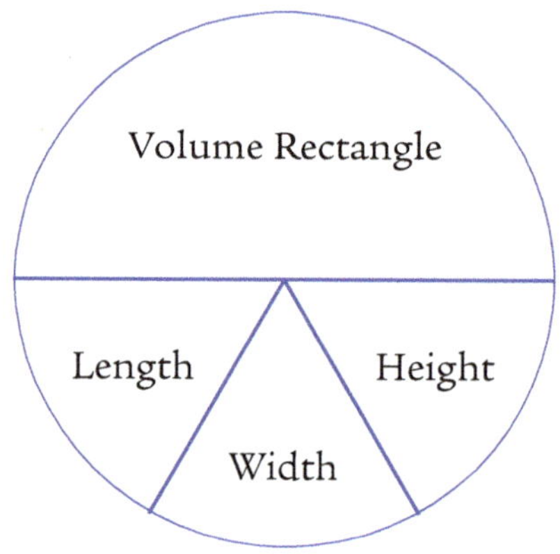

NEED-TO-KNOW CRITERIA AND SAMPLE QUESTIONS

Operate Equipment

	Class I	Class II	Class III	Class IV
Blowers and compressors	Comprehension	Comprehension	Application	Application
Boring equipment	N/A	N/A	Comprehension	Comprehension
Cathodic protection devices	N/A	N/A	Application	Application
Chemical feeders	Comprehension	Comprehension	Comprehension	Application
Cleaning equipment	Comprehension	Application	Analysis	Analysis
Computers	Comprehension	Comprehension	Application	Analysis
Electrical controls	Application	Application	Application	Application
Engines	Application	Application	Application	Application
Excavating equipment	Application	Application	Application	Application
Flow monitoring equipment	Comprehension	Comprehension	Application	Application
Generators	Application	Application	Application	Application
Hand tools	Application	Application	Application	Application
Heavy vehicles	Application	Application	Application	Application
High velocity cleaners	Application	Application	Analysis	Analysis
Inspection equipment (vacuum testing, pressure testing)	Comprehension	Application	Application	Application
Motors	Application	Application	Application	Application
Power tools	Application	Application	Application	Application
Pumps	Application	Application	Application	Application
Rodding equipment	Application	Application	Application	Application
Safety equipment	Application	Application	Application	Application
Tapping equipment	Application	Application	Application	Application
Valves	Application	Application	Application	Application
Variable-speed drives	Application	Application	Application	Application

Required Capabilities

Knowledge of

- Function of tools,
- General chemistry,
- General electrical principles,
- General mechanical principles,
- Hydraulic equipment,
- Instrumentation,
- Internal combustion engines,
- Lubricants and fluids,
- Operation and maintenance practices,
- Physical science,
- Pipe fittings and joining methods,
- Pipeline construction principles,
- Piping material type and size,
- Pneumatics,
- Potential causes of disasters in system,
- Potential impact of disasters on system,
- Regulations,
- Safety regulations,
- Startup and shut-down procedures,
- System operation and maintenance,
- Types of lift stations,
- Types of pumps,
- Wastewater collection system operation and maintenance, and
- Wastewater treatment concepts.

Ability to

- Adjust equipment,
- Calibrate equipment,
- Discriminate between normal and abnormal conditions,
- Evaluate cause of damage,
- Evaluate operation of equipment,
- Follow written procedures,
- Identify cause of damage,
- Identify potential safety hazards,
- Interpret data,

- Interpret material safety data sheets,
- Locate cause of flow problems,
- Maintain system in normal operating condition,
- Monitor electrical equipment,
- Monitor mechanical equipment,
- Operate safety equipment,
- Perform impact assessments,
- Perform mathematical calculations,
- Perform physical measurements,
- Read plans and profiles,
- Recognize unsafe work conditions, and
- Record information.

Class I Sample Questions

1. Which of the following equipment types is most commonly used to excavate sewer trenches?

a) Clamshell
b) Backhoe
c) Front-end loader
d) Trenching machine

2. High-velocity cleaners are primarily used in 4- to 20-in. (102- to 508-mm) diameter pipes for which preventive maintenance purpose?

a) Clear and flush debris
b) Remove roots
c) Remove hardened grease
d) Open blocked pipe

3. A __________ machine can be an effective tool when used to remove sewer stoppages in emergency situations.

a) Pigging
b) Bucket
c) Flushing
d) Rodding

4. Which of the following types of cleaning equipment is most commonly used when responding to an emergency stoppage service request for an overflowing manhole?

   a) Hand rods and power rodders
   b) Hand rods and high-velocity cleaners
   c) Power rodders and bucket machines
   d) High-velocity cleaners and bucket machines

5. Which of the following rodding tools is used to pilot a hole through a pipe experiencing heavy roots, grease, or solids?

   a) Sand corkscrew
   b) Auger
   c) Porcupine
   d) Spring blade cutter

6. The __________ is specified by the manufacturer for a single vehicle, and it must not be exceeded when the vehicle is fully loaded.

   a) Adjusted vehicle weight (AVW)
   b) Gross vehicle weight (GVW)
   c) Adjusted vehicle weight rating (AVWR)
   d) Gross vehicle weight rating (GVWR)

Answers

1. **Answer:** B

   The most commonly used type of equipment for trench excavations is the backhoe.

   Reference: California State University, Sacramento (2003) *Operation and Maintenance of Wastewater Collection Systems*, 6th ed.; California State University: Sacramento, California; Volume I, p 428.

2. **Answer:** A

   High-velocity cleaners are used primarily as a preventive maintenance machine to clear and flush debris such as organic material, grease, and soap/detergent deposits from 4- to 20-in. (102- to 508-mm) diameter pipe.

   Reference: California State University, Sacramento (2003) *Operation and Maintenance of Wastewater Collection Systems*, 6th ed.; California State University: Sacramento, California; Volume I, p 310.

3. **Answer:** D

A rodding machine can be an effective tool when used to remove sewer stoppages during emergency situations.

Reference: California State University, Sacramento (2003) *Operation and Maintenance of Wastewater Collection Systems*, 6th ed.; California State University: Sacramento, California; Volume I, p 292.

4. **Answer:** A

The most commonly used types of equipment when responding to an emergency stoppage service request for an overflowing manhole are hand rods and power rodders. Other equipment such as high-pressure/velocity cleaners are used after the blockage is cleared to remove debris from the line.

Reference: California State University, Sacramento (2003) *Operation and Maintenance of Wastewater Collection Systems*, 6th ed.; California State University: Sacramento, California; Volume I, p 290.

5. **Answer:** B

An auger is used to pilot a hole through a pipe experiencing heavy roots, grease, or solids.

Reference: California State University, Sacramento (2003) *Operation and Maintenance of Wastewater Collection Systems*, 6th ed.; California State University: Sacramento, California; Volume I, p 345.

6. **Answer:** D

The gross vehicle weight rating (GVWR) is specified by the manufacturer for a single vehicle, and it must not be exceeded when the vehicle is fully loaded.

Reference: California State University, Sacramento (2003) *Operation and Maintenance of Wastewater Collection Systems*, 6th ed.; California State University: Sacramento, California; Volume I, p 87.

Class II Sample Questions

1. Which of the following testing methods is used to determine if a newly installed pipe has deflected more than the acceptable design limit?

   a) Air
   b) Hydrostatic
   c) Mandrel
   d) Closed circuit television (CCTV) inspection

2. Which of the following may result from an improperly constructed sewer lateral?

   a) Sewer odors entering a residential building
   b) Excessive infiltration entering the upstream manhole
   c) Corrosion caused by septic conditions within the mainline sewer
   d) A sewer main blockage caused by a protruding tap

3. Horizontal directional drilling and pipe jacking are examples of what type of method?

   a) Trenchless pipe replacement
   b) Open cut pipe installation
   c) Service connection repair
   d) Pipeline renewal

4. Examples of __________ cleaning methods include hand rodding, power rodding, and bucketing.

   a) Hydraulic
   b) Pneumatic
   c) Mechanical
   d) Operational

5. __________ is a method designed to support the walls of a trench and consists of a framework composed of wood and/or metal.

   a) Sheeting
   b) Shielding
   c) Shoring
   d) Sloping

6. Pressure transducers, ultrasonics, and bubblers are used to determine what component of sewer flow?

   a) Depth
   b) Inflow and infiltration
   c) Chemical composition
   d) Temperature

7. Which of the following are common sense procedures to prevent electrocution when working on electrical equipment?

   a) Wear metal-framed eyeglasses, jewelry, or brass belt buckles.
   b) Complete the work quickly, using both hands.

c) Use rubber mats, rubber boots, leather gloves, or other forms of isolation between you and the ground.

d) Before conducting work, test for oxygen deficiency/enrichment and flammable or explosive conditions.

8. Variable-frequency drive (VFD) pumps

- **a)** Do not need maintenance as often as other pumps
- **b)** Are more complex and require specialized skills to maintain
- **c)** Can be used without a motor control center
- **d)** Have chemical battery backup

9. When is it more appropriate to conduct low-pressure air testing on sewers than conducting water tests?

- **a)** In steep terrain
- **b)** During wet weather
- **c)** In areas with high groundwater tables
- **d)** Before sewer laterals are connected to the main line

Answers

1. Answer: C

A mandrel test is used to determine if a newly installed pipe has deflected more than the designer's specified acceptable amount.

Reference: Water Environment Federation (2010) *Wastewater Collection Systems Management*, 6th ed.; Manual of Practice No. 7; Water Environment Federation: Alexandria, Virginia; p 131.

2. Answer: D

An improperly constructed sewer lateral tap may result in a sewer main blockage caused by a protruding tap.

Reference: California State University, Sacramento (2003) *Operation and Maintenance of Wastewater Collection Systems*, 6th ed.; California State University: Sacramento, California; Volume I, pp 52 and 401.

3. Answer: A

Horizontal directional drilling and pipe jacking are examples of trenchless pipe replacement methods.

Reference: Water Environment Federation; American Society of Civil Engineers; Environmental & Water Resources Institute (2009) *Existing Sewer Evaluation and Rehabilitation*, 3rd ed.; WEF

Manual of Practice No. FD-6; ASCE Manuals and Reports on Engineering Practice No. 62; Water Environment Federation: Alexandria, Virginia; pp 226–236.

4. **Answer:** C

Hand rodding, power rodding, and bucketing are examples of mechanical cleaning methods.

Reference: California State University, Sacramento (2003) *Operation and Maintenance of Wastewater Collection Systems*, 6th ed.; California State University: Sacramento, California; Volume I, p 330.

5. **Answer:** C

Shoring is a method designed to support the walls of a trench, and consists of a framework composed of wood and/or metal.

Reference: California State University, Sacramento (2003) *Operation and Maintenance of Wastewater Collection Systems*, 6th ed.; California State University: Sacramento, California; Volume I, p 149.

6. **Answer:** A

Pressure transducers, ultrasonics, and bubblers are used to determine the depth of sewer flow.

Reference: Water Environment Federation (1999) *Prevention and Control of Sewer System Overflows*, 2nd ed.; Manual of Practice No. FD-17; Water Environment Federation: Alexandria, Virginia; p 75.

7. **Answer:** C

The use of rubber mats, rubber boots, leather gloves, or other forms of isolation between you and the ground are common-sense procedures to prevent electrocution when working on electrical equipment.

Reference: California State University, Sacramento (2003) *Operation and Maintenance of Wastewater Collection Systems*, 6th ed.; California State University: Sacramento, California; Volume I, p 155.

8. **Answer:** B

Variable-frequency drive pumps are more complex than other types of pumps and require higher electrical and electronic skill levels.

Reference: California State University, Sacramento (2010) *Operation and Maintenance of Wastewater Collection Systems*, 7th ed.; California State University: Sacramento, California; Volume II, p 34.

9. **Answer:** A

It is more appropriate to conduct low-pressure air testing on sewers in steep terrain than conducting water tests because of the excessive water pressures created at the lower end of a sewer line.

Reference: Water Environment Federation (2010) *Wastewater Collection Systems Management*, 6th ed.; Manual of Practice No. 7; Water Environment Federation: Alexandria, Virginia; p 465.

Class III Sample Questions

1. Supervisory pump controls are typically

   a) Not able to be used in supervisory control and data acquisition (SCADA) management systems
   b) Used to adjust pump operating conditions
   c) Secondary controls taking information from primary controls
   d) Not used in modern design pumping stations

2. A supervisor and two employees are responsible for maintaining a particular lift station and its pumps. If a pump needs to be repaired and must be locked out/tagged out, who should place a lock and a tag on the motor control center (MCC) to secure the pump and its power supply?

   a) The supervisor
   b) The supervisor and one of the employees
   c) Each person who works on the equipment
   d) None of them; maintenance takes care of that

3. To restart a pumping station after maintenance activities, an operator should first

   a) Ensure the pump motor is turning in the proper direction
   b) Notify downstream pumping stations or treatment plants that the station will come online
   c) Stop and start the pump(s) to clear debris that may be present
   d) Open the suction valves or direct flow into the wet well

4. The main benefit of routine maintenance in a sewer system is that it

   a) Eliminates system user complaints
   b) Provides some of the data used in modeling and performance statistics
   c) Eliminates the need for expensive repairs over the long term
   d) Extends the life of the system by keeping hydraulic capacity at design standards

5. A benefit to keeping thorough and organized maintenance records on individual pieces of equipment such as pumps is to __________.

a) Determine future flows through a pumping station
b) Monitor sewers for toxic discharges
c) Ensure employees' whereabouts are tracked during their time on the clock
d) Provide valuable data upon which to base decisions

6. Performing maintenance or repairs after a problem occurs in a piece of equipment is an example of __________ maintenance.

a) Corrective
b) Preventive
c) Predictive
d) Emergency

7. A megohmmeter is used to measure __________.

a) Insulation resistance
b) Heat loss
c) Ground fault
d) Amperage drop

Answers

1. Answer: B

Supervisory controls take information from primary pump start/stop controllers such as bubblers or floats and convert the measurement being taken into a signal to start, stop, or alter speed of the pump(s).

Reference: California State University, Sacramento (2010) *Operation and Maintenance of Wastewater Collection Systems*, 7th ed.; California State University: Sacramento, California; Volume II, p 44.

2. Answer: C

When locking and tagging out equipment for repair, all employees working on the equipment should use their own locks and tags for securing the equipment. This ensures that everyone present knows the equipment is down and in the process of being serviced. It also prevents anyone from accidentally trying to operate that equipment while others are still working.

Reference: Water Environment Federation (2012) *Safety, Health, and Security in Wastewater Systems*, 6th ed.; Manual of Practice No. 1; Water Environment Federation: Alexandria, Virginia; pp 144–145.

3. **Answer:** D

Diverting flow into the wet well or opening suction valves in dry well pumps should be the first step in ensuring the pumps have water before they start.

Reference: California State University, Sacramento (2010) *Operation and Maintenance of Wastewater Collection Systems*, 7th ed.; California State University: Sacramento, California; Volume II, p 92.

4. **Answer:** D

The main benefit of routine maintenance is keeping the hydraulic capacity of the system as high as possible, which lowers overall operation and maintenance costs, extends the useful life of the gravity system, and lowers rehabilitation and replacement costs.

Reference: California State University, Sacramento (2010) *Operation and Maintenance of Wastewater Collection Systems*, 7th ed.; California State University: Sacramento, California; Volume II, p 636.

5. **Answer:** D

Providing maintenance records for individual pieces of equipment is useful to determine the correct frequency of maintenance activities. Documenting maintenance costs over time also can help to justify the need to replace an aging piece of equipment. Without maintenance records, there is no logical documented reason for changing maintenance activity frequency.

Reference: Water Environment Federation (2009) *Wastewater Collection Systems Management*; WEF Manual of Practice MOP-7; Water Environment Federation: Alexandria, Virginia; p 64.

6. **Answer:** A

Corrective maintenance is considered to be any repair or maintenance performed when equipment does not operate normally. Although the maintenance may not be critical at the time of occurrence, it is necessary to correct the problem and to return the equipment to a full operational level.

Reference: California State University, Sacramento (2010) *Operation and Maintenance of Wastewater Collection Systems*, 7th ed.; California State University: Sacramento, California; Volume II, p 585.

7. **Answer:** A

Reference: California State University, Sacramento (2010) *Operation and Maintenance of Wastewater Collection Systems*, 7th ed.; California State University: Sacramento, California; Volume II, p 159.

Class IV Sample Questions

1. Open cut pipeline repair/replacement can often be more economical when

a) The pipe alignment being replaced is less than 10-in. (255-mm) diameter
b) Large-diameter pipes (more than 36-in. [915-mm]) are involved
c) Depth is shallow and soils are stable without groundwater
d) There are few service laterals along the alignment

2. Tracking equipment performance and operating characteristics over time is necessary for __________ maintenance.

a) Emergency
b) Predictive
c) Planned
d) Preventive

3. To provide planning capability for future equipment needs, a(n) __________ should be developed.

a) Needs-based analysis
b) Equipment repair/replacement fund
c) Capital improvement plan
d) Financial stability plan

4. Of the following, which is a valid reason why personal protection equipment (PPE) would not be required?

a) It creates discomfort and/or limits full movement of an individual
b) Costs of obtaining the equipment are high and risk of injury is low
c) Acting as a substitute for engineering, work practices, or administrative control
d) Personnel find an alternative means of accomplishing the same task

5. Hydromechanical cleaning augments large-diameter pipe cleaning using

a) Special nozzles mounted on a cleaning hose, powered by a pump
b) Hydraulic-powered steel rod with large cutting or debris removal tools
c) Air pressure and flotation to remove material at the crown of the pipe
d) Water head to propel the tool downstream and remove debris

6. To best determine the effectiveness of a rehabilitation program, ________ are necessary.

a) Capital improvement project priority lists
b) Decisions on the type of rehabilitation materials to use

c) Funding budgets of 3 to 5 years
d) Measurement, analysis, and documentation

Answers

1. **Answer:** C

Open-cut construction can offer several distinct advantages to other forms of repair/replacement techniques. When the depth of installation is shallow, soils are stable, and no groundwater is present, open-cut construction is often more cost-effective than trenchless options. The open-cut option also becomes more cost-effective when many lateral connections at short intervals need to be reinstated.

Reference: Water Environment Federation; American Society of Civil Engineers; Environmental & Water Resources Institute (2009) *Existing Sewer Evaluation and Rehabilitation*, 3rd ed.; WEF Manual of Practice No. FD-6; ASCE Manuals and Reports on Engineering Practice No. 62; Water Environment Federation: Alexandria, Virginia; p 236.

Additional Reference: California State University, Sacramento (2003) *Operation and Maintenance of Wastewater Collection Systems*, 7th ed.; California State University: Sacramento, California; Volume II, Chapter 10, Section 10.220, pp 284–285.

2. **Answer:** B

Predictive maintenance requires a performance baseline to be established for a system and/or pieces of equipment along with a way to monitor changes in this performance over time. Monitoring performance in this way allows for fairly high accuracy in predicting when equipment or system areas are likely to fail and allows proactive work to occur to avoid the failure.

Reference: California State University, Sacramento (2010) *Operation and Maintenance of Wastewater Collection Systems*, 7th ed.; California State University: Sacramento, California; Volume II, Chapter 13, Section 13.600, p 574.

3. **Answer:** B

To properly prepare for equipment replacement needs, an equipment replacement fund should be developed. This fund should consider all capital equipment used by the organization and calculate the usable life of the equipment to properly fund replacement.

Reference: California State University, Sacramento (2010) *Operation and Maintenance of Wastewater Collection Systems*, 7th ed.; California State University: Sacramento, California; Volume II, Section 14.232, p 612.

4. **Answer:** C

Personal protection equipment should not be used as a substitute for engineering, work practices, or administrative controls. Risk mitigation demands attention to these specific issues. Personal protection equipment only should be used to protect against hazards that are unavoidable.

Reference: California State University, Sacramento (2010) *Operation and Maintenance of Wastewater Collection Systems*, 7th ed.; California State University: Sacramento, California; Volume II, Section 11.83, p 367.

5. **Answer:** D

Hydromechanical cleaning uses water head in the pipe to propel the tool downstream to remove grit and other solids. Sewer balls, kites, and scooters are examples of these devices.

Reference: Water Environment Federation; American Society of Civil Engineers; Environmental & Water Resources Institute (2009) *Existing Sewer Evaluation and Rehabilitation*, 3rd ed.; WEF Manual of Practice No. FD-6; ASCE Manuals and Reports on Engineering Practice No. 62; Water Environment Federation: Alexandria, Virginia; p 198.

6. **Answer:** D

Measurement, analysis, and documentation of rehabilitation projects provide necessary information for benchmarking and comparison with related records in maintenance or treatment costs.

Reference: Water Environment Federation; American Society of Civil Engineers; Environmental & Water Resources Institute (2009) *Existing Sewer Evaluation and Rehabilitation*, 3rd ed.; WEF Manual of Practice No. FD-6; ASCE Manuals and Reports on Engineering Practice No. 62; Water Environment Federation: Alexandria, Virginia; p 303.

Evaluate and Maintain Equipment

	Class I	Class II	Class III	Class IV
Perform maintenance on equipment				
Blowers and compressors	Comprehension	Comprehension	Comprehension	N/A
Calibration of chemical feeders	Comprehension	Comprehension	Comprehension	Comprehension
Cleaning equipment	Comprehension	Comprehension	Comprehension	Comprehension
Electrical controls	Comprehension	Comprehension	Comprehension	Comprehension
Engines	Comprehension	Comprehension	Comprehension	Comprehension
Excavating equipment	Comprehension	Comprehension	Comprehension	Comprehension
Flow monitoring equipment	Comprehension	Comprehension	Analysis	Analysis
Generators	Comprehension	Application	Application	Application
Hand tools	Application	Application	Application	Application
Heavy vehicles	Comprehension	Comprehension	Comprehension	Comprehension
High-velocity cleaners	Comprehension	Comprehension	Analysis	Analysis
Inspection equipment (vacuum testing, pressure testing)	Comprehension	Application	Analysis	Analysis
Motors	Application	Application	Application	Application
Power tools	Application	Application	Application	Application
Pumps	Application	Application	Analysis	Analysis
Rodding equipment	Application	Application	Analysis	Analysis
Safety equipment	Application	Application	Application	Application
Tapping equipment	N/A	N/A	Application	Application
Valves	Comprehension	Comprehension	Comprehension	Comprehension
Variable-speed drives	Comprehension	Comprehension	Analysis	Analysis
Evaluate operation of equipment				
Inspect equipment for abnormal conditions	Analysis	Analysis	Analysis	Analysis
Measure temperature of equipment	Comprehension	Comprehension	Analysis	Analysis
Read charts	Comprehension	Comprehension	Analysis	Analysis
Read gauges	Analysis	Analysis	Analysis	Analysis
Read meters	Analysis	Analysis	Analysis	Analysis
Troubleshoot electrical equipment	Analysis	Analysis	Analysis	Analysis

Required Capabilities

Knowledge of

- Function of tools,
- General electrical principles,
- Instrumentation,
- Internal combustion engines,
- Lubricants and fluids,
- Operation and maintenance practices,
- Physical science,
- Pneumatics,
- Startup and shut-down procedures,
- System operation and maintenance,
- Types of lift stations,
- Types of pumps, and
- Wastewater collection system operation and maintenance.

Ability to

- Adjust equipment,
- Calibrate equipment,
- Diagnose/troubleshoot,
- Differentiate between preventive and corrective maintenance,
- Discriminate between normal and abnormal conditions,
- Evaluate cause of damage,
- Evaluate operation of equipment,
- Evaluate system performance,
- Follow written procedures,
- Identify cause of damage,
- Identify potential safety hazards,
- Interpret data,
- Maintain system in normal operating condition,
- Monitor mechanical equipment,
- Perform general maintenance,
- Perform impact assessments,
- Perform physical measurements,
- Recognize unsafe work conditions, and
- Record information.

Perform Maintenance on Equipment

Class I Sample Questions

1. A pump curve is specific to each individual pump and illustrates the pumped flow compared to the

   a) Viscosity
   b) Total dynamic head
   c) National Electrical Manufacturers Association (NEMA) class
   d) Best efficiency point

2. A swing or ball check valve on the discharge line of a pump is used to

   a) Prevent cavitation during low-flow periods
   b) Serve as an isolation valve during pump maintenance
   c) Eliminate the need for a wye fitting
   d) Provide positive shutoff from the force main when the pump is off

3. Electrical current is measured in

   a) Volts
   b) Kilovolt amperes
   c) Watts
   d) Amps

4. __________ are required to avoid the risk of accidental operation of related equipment while working on an electrical system.

   a) Air relief valves
   b) Fusible links
   c) Lockout/tagout procedures
   d) Confined space entry permits

5. The minimum permissible oxygen level required for a work space is

   a) 19.5%
   b) 21.5%
   c) 80%
   d) 100%

6. __________ is an example of a level sensing technology that does not come in contact with flow during normal operation.

   a) Ultrasonic
   b) Capacitance
   c) Bubbler system
   d) Float ball

7. A variable-frequency drive will regulate pump flow by adjusting the pump's __________.

   a) Total dynamic head
   b) Speed
   c) Pressure
   d) Efficiency

8. When inspecting a pump impeller, you should look for

   a) Paint thickness
   b) Material of construction
   c) Chips or broken tips
   d) Size

Answers

1. **Answer:** B

   Reference: California State University, Sacramento (2010) *Operation and Maintenance of Wastewater Collection Systems*, 7th ed.; California State University: Sacramento, California; Volume II, Section 8.164, p 69.

2. **Answer:** D

   Reference: California State University, Sacramento (2010) *Operation and Maintenance of Wastewater Collection Systems*, 7th ed.; California State University: Sacramento, California; Volume II, Section 8.172, p 79.

3. **Answer:** D

   Reference: California State University, Sacramento (2010) *Operation and Maintenance of Wastewater Collection Systems*, 7th ed.; California State University: Sacramento, California; Volume II, Section 9.11, p 155.

4. **Answer:** C

   Reference: California State University, Sacramento (2010) *Operation and Maintenance of Wastewater Collection Systems*, 7th ed.; California State University: Sacramento, California; Volume II, Section 11.823, p 362.

5. **Answer:** A

Reference: California State University, Sacramento (2010) *Operation and Maintenance of Wastewater Collection Systems*, 7th ed.; California State University: Sacramento, California; Volume II, Section 11.18, p 347.

6. **Answer:** A

Reference: California State University, Sacramento (2010) *Operation and Maintenance of Wastewater Collection Systems*, 7th ed.; California State University: Sacramento, California; Volume II, Section 8.152, Figure 8.31, pp 49–52.

7. **Answer:** B

Reference: California State University, Sacramento (2010) *Operation and Maintenance of Wastewater Collection Systems*, 7th ed.; California State University: Sacramento, California; Volume II, Section 8.135, p 35.

8. **Answer:** C

Reference: California State University, Sacramento (2010) *Operation and Maintenance of Wastewater Collection Systems*, 7th ed.; California State University: Sacramento, California; Volume II, p 214.

Class II Sample Questions

1. When monitoring flow in an open channel that experiences surcharge conditions, what type of flow monitoring device will provide the most accurate measurements?

   a) Inline flow meter
   b) Area-velocity
   c) Parhall flume and a depth pressure sensor
   d) Magnetic flow meter

2. Variable-frequency drives are used in pumping stations to

   a) Keep the pump revolutions per minute (rpm) at a constant rate
   b) Match the discharge rate to the influent rate
   c) Vary the wet well level to prevent buildup
   d) Turn the pumps on and off frequently

3. When using a hydraulic cleaning machine, the most effective cleaning occurs with

   a) High pressure and fast reel speed
   b) Low pressure and moderate reel speed
   c) Low pressure and fast reel speed
   d) High pressure and moderate reel speed

4. The type of valve used to prevent water from flowing backwards through a pump is called a(n) __________ valve.

   a) Anti-siphon
   b) Gate
   c) Check
   d) Ball

5. The type of motor used in a wet well-only pumping station is called a(n) ________ motor.

   a) Squirrel cage
   b) Wound rotor
   c) Induction
   d) Submersible

6. Face shields, gloves, and steel-toed work shoes are known as __________.

   a) Personal protection equipment
   b) Anti-bacterial protection equipment
   c) Wastewater-resistant equipment
   d) Sewer safety uniform

7. The bottom of the inside of a sewer pipe is called the __________.

   a) Channel
   b) Crown
   c) Invert
   d) Segment

8. Regular cleaning of areas in the sewer system that have frequent blockages is called __________ maintenance.

   a) Predictive
   b) Reactive
   c) Emergency
   d) Preventive

9. One thousand parts per million (ppm) of hydrogen sulfide gas is classified as ________.

   a) Lower explosive limit (LEL)
   b) Immediately dangerous to life or health (IDLH)
   c) Upper explosive limit (UEL)
   d) Time weighted average (TWA)

10. A large pumping station that has separate chambers for the wastewater and the pumps is called a __________ pumping station.

- **a)** Submersible
- **b)** Pneumatic ejector
- **c)** Suction lift
- **d)** Dry well

11. Inverted siphons are used to go under obstructions such as

- **a)** Rivers, highways, and utilities
- **b)** Tree roots and poor soil
- **c)** Mountains and river gorges
- **d)** Private property and groundwater tables

Answers

1. Answer: C

Reference: California State University, Sacramento (2010) *Operation and Maintenance of Wastewater Collection Systems*, 7th ed.; California State University: Sacramento, California; Volume II, Section 10.12, p 276.

2. Answer: B

Reference: California State University, Sacramento (2010) *Operation and Maintenance of Wastewater Collection Systems*, 7th ed.; California State University: Sacramento, California; Volume II, Section 8.135, p 35.

3. Answer: D

Reference: California State University, Sacramento (2003) *Operation and Maintenance of Wastewater Collection Systems*, 6th ed.; California State University: Sacramento, California; Volume I, Section 6.112, p 317.

4. Answer: C

Reference: California State University, Sacramento (2010) *Operation and Maintenance of Wastewater Collection Systems*, 7th ed.; California State University: Sacramento, California; Volume II, Section 8.172, p 78.

5. Answer: D

Reference: California State University, Sacramento (2010) *Operation and Maintenance of Wastewater Collection Systems*, 7th ed.; California State University: Sacramento, California; Volume II, Section 9.315, p 207.

6. **Answer:** A

Reference: Water Environment Federation (2012) *Safety, Health, and Security in Wastewater Systems*, 6th ed.; Manual of Practice No. 1; Water Environment Federation: Alexandria, Virginia; Chapter 10.

7. **Answer:** C

Reference: California State University, Sacramento (2003) *Operation and Maintenance of Wastewater Collection Systems*, 6th ed.; California State University: Sacramento, California; Volume I, Section 3.41, p 60.

8. **Answer:** D

Reference: California State University, Sacramento (2010) *Operation and Maintenance of Wastewater Collection Systems*, 7th ed.; California State University: Sacramento, California; Volume II, Section 13.600, p 574.

9. **Answer:** B

Reference: California State University, Sacramento (2003) *Operation and Maintenance of Wastewater Collection Systems*, 6th ed.; California State University: Sacramento, California; Volume I, Section 4.41, p 118.

10. **Answer:** D

Reference: California State University, Sacramento (2010) *Operation and Maintenance of Wastewater Collection Systems*, 7th ed.; California State University: Sacramento, California; Volume II, Section 8.012, p 14.

11. **Answer:** A

Reference: Water Environment Federation (2010) *Wastewater Collection Systems Management*, 6th ed.; Manual of Practice No. 7; Water Environment Federation: Alexandria, Virginia; p 132.

Class III Sample Questions

1. Two types of alternating current induction motor construction encountered in wastewater pumps include

   a) Submersible and diaphragm
   b) Diaphragm and static
   c) Variable and diaphragm
   d) Squirrel cage and wound

2. Screw pumps consist of a screw operating at a __________ speed.

   a) Variable
   b) High
   c) Constant
   d) Low

3. An ejector pump will not discharge if __________.

   a) There is insufficient air pressure or volume
   b) The discharge gate is open
   c) The counterweight is too far from the glide
   d) Pressure in the upstream sewer is too low

4. Piston pumps are commonly used in

   a) Sumps
   b) Wet wells
   c) High-velocity cleaners
   d) Manholes

5. The air tank on a compressor should be drained of condensate on a __________ basis.

   a) Daily
   b) Weekly
   c) Monthly
   d) Semiannually

6. Before starting to repack a pump, you should know

   a) The size of the shaft
   b) The number of impellers needed
   c) Where to position the seal cage (lantern ring)
   d) The size of the pump head

7. Compressor crankcase reservoirs should be inspected

   a) Daily
   b) Monthly
   c) Semiannually
   d) Yearly

8. Air compressors can be used to operate

a) Valve operators
b) Scissor lift
c) Generators
d) Centrifugal pumps

9. Fuses are electrical devices that __________ circuits when the current flowing through them exceeds the rating of the fuse.

a) Close
b) Open
c) Load
d) Energize

10. The ammeter measures the __________ in the circuit.

a) Voltage
b) Current
c) Phase
d) Insulation resistance

11. Effective grounding means that the path to ground is

a) Permanent and continuous
b) Not accessible
c) Three phased
d) Thermal magnetically protected

12. Typically, motors up to 100 hp should not start more than __________ times per hour.

a) 4 or 5
b) 7 or 8
c) 10 or 11
d) 13 or 14

Answers

1. Answer: D

Reference: California State University, Sacramento (2010) *Operation and Maintenance of Wastewater Collection Systems*, 7th ed.; California State University: Sacramento, California; Volume II, p 167.

2. **Answer:** C

Reference: California State University, Sacramento (2010) *Operation and Maintenance of Wastewater Collection Systems*, 7th ed.; California State University: Sacramento, California; Volume II, p 202.

3. **Answer:** A

Reference: California State University, Sacramento (2010) *Operation and Maintenance of Wastewater Collection Systems*, 7th ed.; California State University: Sacramento, California; Volume II, p 204.

4. **Answer:** C

Reference: California State University, Sacramento (2010) *Operation and Maintenance of Wastewater Collection Systems*, 7th ed.; California State University: Sacramento, California; Volume II, p 205.

5. **Answer:** A

The air tank on a compressor should be drained at least daily to remove the condensate that has accumulated in the tank.

Reference: California State University, Sacramento (2010) *Operation and Maintenance of Wastewater Collection Systems*, 7th ed.; California State University: Sacramento, California; Volume II, p 250.

6. **Answer:** C

Reference: California State University, Sacramento (2010) *Operation and Maintenance of Wastewater Collection Systems*, 7th ed.; California State University: Sacramento, California; Volume II, p 219.

7. **Answer:** C

Reference: California State University, Sacramento (2010) *Operation and Maintenance of Wastewater Collection Systems*, 7th ed.; California State University: Sacramento, California; Volume II, p 249.

8. **Answer:** A

Reference: California State University, Sacramento (2010) *Operation and Maintenance of Wastewater Collection Systems*, 7th ed.; California State University: Sacramento, California; Volume II, p 249.

9. **Answer:** B

Reference: California State University, Sacramento (2010) *Operation and Maintenance of Wastewater Collection Systems*, 7th ed.; California State University: Sacramento, California; Volume II, p 162.

10. Answer: B

Reference: California State University, Sacramento (2010) *Operation and Maintenance of Wastewater Collection Systems*, 7th ed.; California State University: Sacramento, California; Volume II, p 158.

11. Answer: A

Reference: California State University, Sacramento (2010) *Operation and Maintenance of Wastewater Collection Systems*, 7th ed.; California State University: Sacramento, California; Volume II, p 165.

12. Answer: A

Reference: California State University, Sacramento (2010) *Operation and Maintenance of Wastewater Collection Systems*, 7th ed.; California State University: Sacramento, California; Volume II, p 171.

Class IV Sample Questions

1. A low-pressure air test was conducted on a new 280-ft (85-m) long section of 8-in. (203-mm) polyvinylchloride (PVC) pipe. The test identifies a vapor leak within the new section. This result indicates

- **a)** The new section of pipe was tested under too high a pressure.
- **b)** The groundwater table is flowing below the pipe causing a siphon effect in the bedding.
- **c)** The test pressure gauge is not reading accurately.
- **d)** A small crack or poorly sealed joint is allowing air to escape.

2. The pressure gauge on the discharge side of a pump experiences a drop in line pressure from 6 psi (41 kPa) down to 0 psi (0 kPa) between cycles. What is the most likely explanation for the loss of pressure?

- **a)** The discharge gate valve is not holding its seal and may need to be cleaned or repaired.
- **b)** The suction gate valve is not holding its seal and may need to be cleaned or repaired.
- **c)** The check valve is not holding its seal and may need to be cleaned or replaced.
- **d)** The vacuum break or air release valve needs to be flushed and cleaned.

3. A recent camera inspection reveals a 50% obstruction caused by grease. What nozzle should the combination truck operator select to remove the grease with a high-velocity cleaner?

- **a)** Crown obstruction nozzle
- **b)** Preventive maintenance radial nozzle

c) Skid-mounted sand nozzle
d) Tear drop nozzle

4. When performing excavation work for either a new pipe installation or repair in a Type A cohesive soil, at what depth does a trench require shoring?

a) 2 ft (0.6 m)
b) 5 ft (1.5 m)
c) 8 ft (2.4 m)
d) 10 ft (3.0 m)

5. In a traffic control zone, where should equipment and construction materials be stored?

a) Advanced warning area
b) Transition area
c) Buffer area
d) Work area

6. When calculating brake horsepower (bhp), which head value should be considered?

a) Total static head
b) Static negative suction head
c) Total dynamic head
d) Static discharge head

7. While attempting to clear a blockage in a 12-in. (305-mm) line with a sectional rod, the spearhead tool has broken off approximately 75 ft (23 m) from the manhole because of over-torque. What tool must be used to retrieve the broken rod?

a) Pickup tool
b) Auger
c) Square bar corkscrew
d) Pullout tool

8. When working in a manhole, what is the proper way to provide hand tools to the entrant that is conducting the repair?

a) Drop the tool down for the entrant to catch.
b) Have the entrant climb up the ladder to take the tool by hand off.
c) Lower the tools into the manhole in a bucket before the person has entered.
d) Lower the tools into the manhole in a bucket after the person has entered.

9. When connecting a motor to a pump, what are the two potential forms of coupling misalignment that must be avoided to ensure proper operation?

   a) Angular and parallel
   b) Nonlinear and linear
   c) Severe and moderate
   d) End float and nonlinear

10. When filling the water tank on a high-velocity cleaning truck from a fire hydrant, the operator must always remember to __________.

   a) Connect a hose directly to the hydrant and use a pipe wrench to open the hydrant valve halfway to control the flow to the tank.
   b) Use an air gap device, connect a hose and gate valve, use a hydrant wrench to slowly open the hydrant valve all the way, and flush the hydrant before filling the tank.
   c) Use an air gap device, connect a hose and gate valve, use a hydrant wrench to quickly open the hydrant valve all the way, and flush the hydrant before filling the tank.
   d) Use an air gap device, connect a hose and gate valve, and use a pipe wrench to throttle the hydrant valve to control the flow to the tank.

11. After locking out and isolating all of the potential hazardous energy sources on a pump and control panel, the operator must __________.

   a) Immediately expedite the repair before the lock is removed by the attendant.
   b) Ask the supervisor if they are sure that the energy is off.
   c) Remove the fuse to ensure that the energy is off.
   d) Attempt to run the equipment using the controls to ensure that the energy is off.

Answers

1. **Answer:** D

The vapor leak indicates a small crack or poorly sealed joint is allowing the compressed air to escape from the new section, which could attract roots and compromise the pipe or introduce infiltration.

Reference: California State University, Sacramento (2003) *Operation and Maintenance of Wastewater Collection Systems*, 6th ed.; California State University: Sacramento, California; Volume I, p 465.

2. **Answer:** C

Typically, a check valve is installed in the discharge of a pump to provide a positive shutoff from force main pressure so that, when the pump is shut off, the force main will be prevented from draining back into the wet well.

Reference: California State University, Sacramento (2010) *Operation and Maintenance of Wastewater Collection Systems*, 7th ed.; California State University: Sacramento, California; Volume II, p 78.

3. **Answer:** A

Because grease typically adheres to the crown of the pipe, a crown obstruction nozzle designed to remove grease, waste oils, and detergents from the top of the pipe would be best suited for the job.

Reference: California State University, Sacramento (2003) *Operation and Maintenance of Wastewater Collection Systems*, 6th ed.; California State University: Sacramento, California; Volume I, p 316.

4. **Answer:** B

All trenches 5 ft in depth or deeper must be shored to protect operators that are working in the trench.

Reference: California State University, Sacramento (2003) *Operation and Maintenance of Wastewater Collection Systems*, 6th ed.; California State University: Sacramento, California; Volume I, p 408.

5. **Answer:** D

The work area is the area where the work activity is taking place and this is where workers, equipment, and material should be staged. Nothing should ever be stored in the buffer zone. This space, located ahead of the work zone, is dedicated to allowing a shadow truck or an out-of-control motorist the opportunity to recover before entering the work zone.

Reference: California State University, Sacramento (2003) *Operation and Maintenance of Wastewater Collection Systems*, 6th ed.; California State University: Sacramento, California; Volume I, p 96.

6. **Answer:** C

Total dynamic head (TDH) is the measurement of the amount of energy that a pump must develop to move the liquid being pumped. This value is, therefore, used to calculate brake horsepower. Total dynamic head takes into consideration total static head, velocity head, and all of the minor losses experienced across the pump.

Reference: California State University, Sacramento (2010) *Operation and Maintenance of Wastewater Collection Systems*, 7th ed.; California State University: Sacramento, California; Volume II, p 68.

7. Answer: A

A pickup tool turns slowly in the sewer and locks onto the rod coupling, allowing the string of broken rod to be removed safely. The broken rod must be removed or it will be the cause of a future blockage as debris catches on it and accumulates in the pipe.

Reference: California State University, Sacramento (2003) *Operation and Maintenance of Wastewater Collection Systems*, 6th ed.; California State University: Sacramento, California; Volume I, p 347.

8. Answer: C

Tools should be lowered into and pulled out of manholes in a bucket before the operator enters and after the entrant has exited the manhole. If tools must be lowered into a manhole that is occupied by an entrant, then the tools should be secured inside the bucket or bag in a manner that would ensure that the tools do not fall out of the bucket while being lowered and strike the entrant, causing injury.

Reference: California State University, Sacramento (2003) *Operation and Maintenance of Wastewater Collection Systems*, 6th ed.; California State University: Sacramento, California; Volume I, p 120.

9. Answer: A

When connecting a pump and a motor, there are two important types of misalignment: parallel and angular.

Reference: California State University, Sacramento (2010) *Operation and Maintenance of Wastewater Collection Systems*, 7th ed.; California State University: Sacramento, California; Volume II, p 179.

10. Answer: B

When filling the water tank on a high-velocity cleaning truck from a fire hydrant, the operator must use an air gap device to protect the water supply from contamination. The hose should be configured with a gate valve to provide control of the feed into the tank. It is important to slowly open the hydrant valve all the way using only a hydrant wrench. Using the hydrant valve to throttle the feed to the truck could damage the hydrant or cause a hammer condition in the water system. The hydrant should always be flushed before filling the truck's reservoir to prevent the introduction of particles, which may inhibit the pump's performance.

Reference: California State University, Sacramento (2003) *Operation and Maintenance of Wastewater Collection Systems*, 6th ed.; California State University: Sacramento, California; Volume I, pp 315, 319, 321.

11. Answer: D

After locking out and isolating all of the potential hazardous energy sources on a pump and control panel, the operator must try to run the equipment using the locked out controls to ensure that the energy is off.

Reference: California State University, Sacramento (2003) *Operation and Maintenance of Wastewater Collection Systems*, 6th ed.; California State University: Sacramento, California; Volume I, p 159.

Evaluate Operation of Equipment

Class I Sample Questions

1. Which of the following situations may result if there is too much distance between a hand rod power drive and the manhole rod guide tube during hand rodding operation?

   a) A loop can form in the rod, the rod can break, and someone can get injured by the broken rod.
   b) A sudden surge of stored wastewater can shoot through the rod tube when a stoppage is broken loose.
   c) The rod can be pushed through the pipe too quickly, and roots and other material may remain in the pipe.
   d) The rod can seize in the palm of the operator's hand and cause severe injury.

2. When reading an atmospheric analyzer, the operator should make sure the oxygen percentage reading is between __________.

   a) 12.6 and 15.9%
   b) 16.2 and 19.1%
   c) 19.5 and 23.5%
   d) 24.5 and 26.5%

Answers

1. Answer: A

If there is too much distance between a hand rod power drive and the manhole rod guide tube during hand rodding operation, a loop can form in the rod, the rod can break, and someone can get injured by the broken rod.

Reference: California State University, Sacramento (2003) *Operation and Maintenance of Wastewater Collection Systems*, 6th ed.; California State University: Sacramento, California; Volume I, p 358.

2. **Answer:** C

Reference: California State University, Sacramento (2003) *Operation and Maintenance of Wastewater Collection Systems*, 6th ed.; California State University: Sacramento, California; Volume I, Section 4.520, p 128.

Class II Sample Questions

1. Emergency maintenance of lift stations is less desirable than other types of maintenance because

   a) Different personnel need to perform emergency maintenance because of labor union rules.
   b) There are potential regulatory consequences and the likelihood of substantially higher costs.
   c) Personnel will have higher morale.
   d) Bypass pumping equipment is required for emergency maintenance only.

2. Force main air valves should be routinely checked for operation because

   a) Air constantly comes out of solution and needs to be relieved.
   b) Air valve access chambers require periodic cleaning.
   c) Force mains typically require resizing to accommodate different flows.
   d) Air valves control air flow into the pumps.

3. To detect early signs of trouble related to pump performance, periodic cleaning and a simple __________ should be performed.

   a) Ultrasonic test
   b) Vibration analysis
   c) Look-feel inspection
   d) Oil analysis

Answers

1. **Answer:** B

Emergency maintenance is typically far most costly to conduct than preventive, corrective, or predictive maintenance. Additionally, pumping station failures can have a negative effect on the environment, and regulatory fines or other actions may result.

Reference: Water Environment Federation (2010) *Wastewater Collection Systems Management*, 6th ed.; Manual of Practice No. 7; Water Environment Federation, Alexandria, Virginia; p 37.

2. **Answer:** A

Air enters pressure pipe (force mains) from pumps and comes out of solution as wastewater is subjected to changing pressures and velocities. If this air is not relieved, pressure can build over time and rupture the force main or cause a flow interruption that can damage pumps and other system components.

Reference: Water Environment Federation (2010) *Wastewater Collection Systems Management*, 6th ed.; Manual of Practice No. 7; Water Environment Federation: Alexandria, Virginia; p 118.

3. **Answer:** C

To detect early signs of trouble related to pump performance, periodic cleaning and a simple "look-feel" inspection should be performed. Look-feel inspections include observing pumps during operation; checking pump discharge lines, pump seal water, and control panel switches; and monitoring discharge flowrates, pump revolutions per minute, and pump pressures.

Reference: Water Environment Federation (2007) *Operation of Municipal Wastewater Treatment Plants*; Manual of Practice No. 11, 6th ed.; Water Environment Federation: Alexandria, Virginia; pp 8-24.

Class III Sample Questions

1. After a lift station has been constructed, it is inspected by the contractor, engineer, and __________ before final acceptance.

    a) Local building inspector
    b) Operating agency
    c) State environmental office
    d) Electrical inspector

2. Infrared scanning technology in pumping stations is useful in assessing

    a) Cracks in pump housings, shafts, and impellers
    b) Small vibrational movements in mountings and shaft operation
    c) Temperature variances in electrical systems, motor starters, and circuit breakers
    d) Leaks in system components

Answers

1. **Answer:** B

Reference: California State University, Sacramento (2010) *Operation and Maintenance of Wastewater Collection Systems*, 7th ed.; California State University: Sacramento, California; Volume II, p 82.

2. **Answer:** C

Infrared scanning detects temperature variances in electrical boxes, circuit breakers, motors and motor starters, overload relays, fuses and holders, and capacitors. Infrared scanning is a very functional and effective predictive maintenance tool in pumping stations that can be used to help identify problems before they develop into corrective or emergency repairs.

Reference: Water Environment Federation (2010) *Wastewater Collection Systems Management*, 6th ed.; Manual of Practice No. 7; Water Environment Federation: Alexandria, Virginia; p 37.

Class IV Sample Questions

1. Which of the following diagnostic techniques can be used for early detection of mechanical problems related to rotating components such as bent shafts, defects in bearings, and impeller imbalance?

   a) Vibration analysis
   b) Electroflux check
   c) Draw-down test
   d) Air test

2. Multimeters are used to troubleshoot electrical equipment. Multimeters can be used to measure voltage, current, and __________.

   a) Ohms
   b) Flow
   c) Resistance
   d) Horsepower

Answers

1. **Answer:** A

Reference: Water Environment Federation (2010) *Wastewater Collection Systems Management*, 6th ed.; Manual of Practice No. 7; Water Environment Federation: Alexandria, Virginia; pp 37–39.

2. Answer: C

Reference: California State University, Sacramento (2010) *Operation and Maintenance of Wastewater Collection Systems*, 7th ed.; California State University: Sacramento, California; Volume II, p 156.

Maintain and Restore Collection Systems

	Class I	Class II	Class III	Class IV
Clean system				
Hydraulic cleaning (balling, flushing, poly pig)	Application	Application	Application	Analysis
Jet rodding	Application	Application	Application	Analysis
Remove stoppage	Application	Application	Analysis	Analysis
Rodding	Application	Application	Application	Analysis
Root control	Application	Application	Analysis	Analysis
Inspect system				
Dye testing	N/A	Application	Application	Application
Physical inspection	Application	Application	Application	Application
Televising	Comprehension	Comprehension	Analysis	Analysis
Rehabilitate and repair collection system				
Lift station fitting and piping	Application	Application	Analysis	Analysis
Manholes	Application	Application	Analysis	Analysis
Sewer lines	Application	Application	Analysis	Analysis
Taps	Application	Application	Analysis	Analysis

Required Capabilities

Knowledge of

- Function of tools,
- General hydraulic principles,
- Instrumentation,
- Operation and maintenance practices,
- Physical science,
- Pipe fittings and joining methods,
- Pipeline construction principles,
- Piping material type and size,
- Potential causes of disasters in system,
- Potential effect of disasters on system,

- Regulations,
- Risk management,
- Startup and shut-down procedures,
- System operation and maintenance,
- Types of lift stations,
- Types of pumps,
- Wastewater collection design parameters,
- Wastewater collection system concepts,
- Wastewater collection system operation and maintenance, and
- Wastewater treatment concepts.

Ability to

- Adjust equipment,
- Diagnose/troubleshoot,
- Differentiate between preventive and corrective maintenance,
- Discriminate between normal and abnormal conditions,
- Evaluate cause of damage,
- Evaluate data from inspections,
- Evaluate system performance,
- Follow written procedures,
- Identify cause of damage,
- Identify potential safety hazards,
- Interpret data,
- Interpret material safety data sheets,
- Locate cause of flow problems,
- Maintain system in normal operating condition,
- Monitor mechanical equipment,
- Perform general maintenance,
- Perform impact assessments,
- Perform physical measurements,
- Read plans and profiles,
- Recognize unsafe work conditions, and
- Record information.

Class I Sample Questions

1. Smoke testing should be performed on a sewer pipe

a) When groundwater is high to locate broken or leaking pipes
b) On a windy day so the public will not call the fire department

- **c)** At a lift station to find an electrical defect
- **d)** When groundwater is low to find defects

2. Using a closed circuit television (CCTV) camera before removing roots or debris with a hydraulic cleaning nozzle can reveal what serious problem?

- **a)** Pipeline has corrosion or structural flaws.
- **b)** Service taps directly across from each other in a mainline pipe.
- **c)** The sewer main has an internal drop into the manhole.
- **d)** The downstream pipe size is smaller than the pipe to be cleaned.

3. A capacity, management, operations, and maintenance (CMOM) program will

- **a)** Prevent all sanitary sewer overflows (SSOs).
- **b)** Improve the performance of the collection system.
- **c)** Ensure a utility can increase rates for service.
- **d)** Comply with the U.S. Environmental Protection Agency (U.S. EPA) regulation.

4. Which one of the following is not a basic pipe material commercially available for gravity sewer construction?

- **a)** Ductile iron
- **b)** Copper
- **c)** Polyethylene
- **d)** Polyvinyl chloride

Answers

1. Answer: D

Reference: Water Environment Federation; American Society of Civil Engineers; Environmental & Water Resources Institute (2009) *Existing Sewer Evaluation and Rehabilitation*, 3rd ed.; WEF Manual of Practice No. FD-6; ASCE Manuals and Reports on Engineering Practice No. 62; Water Environment Federation: Alexandria, Virginia; p 11.

2. Answer: A

Reference: Water Environment Federation; American Society of Civil Engineers; Environmental & Water Resources Institute (2009) *Existing Sewer Evaluation and Rehabilitation*, 3rd ed.; WEF Manual of Practice No. FD-6; ASCE Manuals and Reports on Engineering Practice No. 62; Water Environment Federation: Alexandria, Virginia; p 9.

3. Answer: B

Reference: Water Environment Federation (2010) *Wastewater Collection Systems Management*, 6th ed.; Manual of Practice No. 7; Water Environment Federation: Alexandria, Virginia; p 10.

Additional reference: California State University, Sacramento (2003) *Operation and Maintenance of Wastewater Collection Systems*, 6th ed.; California State University: Sacramento, California; Volume I, Chapter 2, Section 2.30, p 18.

4. **Answer:** B

Reference: Water Environment Federation (2010) *Wastewater Collection Systems Management*, 6th ed.; Manual of Practice No. 7; Water Environment Federation: Alexandria, Virginia; p 112.

Additional reference: California State University, Sacramento (2003) *Operation and Maintenance of Wastewater Collection Systems*, 6th ed.; California State University: Sacramento, California; Volume I, Chapter 3, Section 3.46, pp 64–65.

Class II Sample Questions

1. Which method is used for manhole renewal in structures that are unsound?

   a) Fiberglass reinforced plastic pipe
   b) High-density polyethylene sheets
   c) Spiral-wound in situ pipe
   d) Cured-in-place liner

2. If the public sewer main is experiencing a sanitary sewer overlow (SSO), then corrective action needs to be taken. Any materials discovered in the line should be removed from the __________ if possible.

   a) Downstream manhole
   b) Upstream manhole
   c) Bar screen at the treatment plant
   d) Nearest pumping station

3. Identify the following related activity that can be combined with predictive, scheduled, and corrective maintenance at a lift station:

   a) Supervisory control and data acquisition
   b) Work management
   c) Risk analysis
   d) Preventive

Answers

1. **Answer:** D

Reference: Water Environment Federation; American Society of Civil Engineers; Environmental & Water Resources Institute (2009) *Existing Sewer Evaluation and Rehabilitation*, 3rd ed.; WEF

Manual of Practice No. FD-6; ASCE Manuals and Reports on Engineering Practice No. 62; Water Environment Federation: Alexandria, Virginia; p 39.

2. **Answer:** A

Reference: Water Environment Federation (2010) *Wastewater Collection Systems Management*, 6th ed.; Manual of Practice No. 7; Water Environment Federation: Alexandria, Virginia; pp 31–32.

3. **Answer:** D

Reference: Water Environment Federation (2010) *Wastewater Collection Systems Management*, 6th ed.; Manual of Practice No. 7; Water Environment Federation: Alexandria, Virginia; p 40.

Class III Sample Questions

1. What is the slope of a 10-in. (254-mm) sewer pipe that drops 6 in. (152 mm) over a distance of 150 ft (46 m)?

    a) 0.3%
    b) 0.6%
    c) 1.5%
    d) 25%

2. A lateral service connection can be restored when using cured-in-place pipe (CIPP) in the sewer main line by the following method:

    a) Laser profiling of the pump
    b) Removing an air plug in the pipe
    c) Cutting the device at the dimples in the pipe
    d) High-pressure nozzle

3. Operations budgets are typically calculated on the basis of resources needed for each of the following four areas: preventive maintenance, inspections, complaint servicing, and __________.

    a) Human resources
    b) Accountant fees
    c) Spot repairs
    d) Board meetings

4. What are the three common factors that are important to prevent accidents?

    a) Work schedule, assignments, and tools
    b) Awareness, responsibility, and commitment

c) Lighting, signage, and staff
d) Equipment, safety gear, and supervision

Answers

1. **Answer:** A

   Slope equals the drop (feet or meters) divided by distance (feet or meters) × 100 = (0.5/150) × 100% = 0.30%

   Reference: Water Environment Federation (2010) *Wastewater Collection Systems Management*, 6th ed.; Manual of Practice No. 7; Water Environment Federation: Alexandria, Virginia; p 114.

2. **Answer:** C

   Reference: California State University, Sacramento (2003) *Operation and Maintenance of Wastewater Collection Systems*, 6th ed.; California State University: Sacramento, California; Volume I, p 310.

3. **Answer:** C

   Reference: Water Environment Federation (2010) *Wastewater Collection Systems Management*, 6th ed.; Manual of Practice No. 7; Water Environment Federation: Alexandria, Virginia; p 51.

4. **Answer:** B

   California State University, Sacramento (2003) *Operation and Maintenance of Wastewater Collection Systems*, 6th ed.; California State University: Sacramento, California; Volume II, p 341.

Class IV Sample Questions

1. A disadvantage of using high-density polyethylene (HDPE) pipe is that

   a) Joint gasket material can deteriorate and leak over time.
   b) The HDPE pipe can expand or contract when pulled through a host pipe.
   c) The HDPE pipe only can be used in a high-pressure, force main location.
   d) Toxic chemicals are used to weld pipe sections together.

2. Which three of these would you gain from a successful preventive maintenance program?

   a) Reduced overtime costs, increased capital repair or replacement costs, improved morale
   b) Reduced materials cost, improved morale, reduced overtime costs
   c) Improved morale, increased capital repair or replacement costs, reduced overtime costs
   d) Improved morale, reduced overtime, reduced materials cost

3. Vibration analysis can detect a variety of electrical and mechanical problems; which of the following groups are all detectable by vibration analysis?

   a) Bent shafts, mechanical loosening, rotation unbalance
   b) Mechanical loosening, bent shafts, overheating
   c) Overheating, bent shafts, rotation unbalance
   d) Rotation unbalance, bent shafts, overheating

Answers

1. **Answer:** B

   Reference: Water Environment Federation; American Society of Civil Engineers; Environmental & Water Resources Institute (2009) *Existing Sewer Evaluation and Rehabilitation*, 3rd ed.; WEF Manual of Practice No. FD-6; ASCE Manuals and Reports on Engineering Practice No. 62; Water Environment Federation: Alexandria, Virginia; p 13.

2. **Answer:** D

   Reference: Water Environment Federation (2010) *Wastewater Collection Systems Management*, 6th ed.; Manual of Practice No. 7; Water Environment Federation: Alexandria, Virginia; pp 38–39.

3. **Answer:** A

   Reference: Water Environment Federation (2010) *Wastewater Collection Systems Management*, 6th ed.; Manual of Practice No. 7; Water Environment Federation: Alexandria, Virginia; p 39.

Maintain Lift Stations

	Class I	Class II	Class III	Class IV
Electrical				
Fuses	Application	Application	Analysis	Analysis
Motors	Application	Application	Analysis	Analysis
Relays	Application	Application	Analysis	Analysis
Starters	Application	Application	Analysis	Analysis
Electronic				
Alarms	Application	Application	Analysis	Analysis
Controllers	Application	Application	Analysis	Analysis
Gas detection	Application	Application	Analysis	Analysis
Level detection system	Application	Application	Analysis	Analysis
Remote terminal units	Application	Application	Analysis	Analysis
Mechanical				
Piping	Application	Application	Analysis	Analysis
Pressure relief valves	Application	Application	Analysis	Analysis
Pretreatment	N/A	N/A	Analysis	Analysis
Pumps	Application	Application	Analysis	Analysis
Valves	Application	Application	Analysis	Analysis
Wet wells	Application	Application	Analysis	Analysis

Required Capabilities

Knowledge of

- Computer operations,
- Function of tools,
- General electrical principles,
- Hazardous situations,
- Instrumentation,
- Internal combustion engines,

- Lubricants and fluids,
- Physical science,
- Pipe fittings and joining methods,
- Pipeline construction principles,
- Piping material type and size,
- Pneumatics,
- Potential causes of disasters in a system,
- Potential effect of disasters on a system,
- Startup and shut-down procedures,
- System operation and maintenance,
- Types of lift stations,
- Types of pumps, and
- Wastewater collection system operation and maintenance.

Ability to

- Adjust equipment,
- Calibrate equipment,
- Diagnose/troubleshoot,
- Differentiate between preventive and corrective maintenance,
- Discriminate between normal and abnormal conditions,
- Evaluate cause of damage,
- Evaluate operation of equipment,
- Evaluate system performance,
- Follow written procedures,
- Identify cause of damage,
- Identify potential safety hazards,
- Interpret data,
- Locate cause of flow problems,
- Maintain system in normal operating condition,
- Monitor electrical equipment,
- Monitor mechanical equipment,
- Operate safety equipment,
- Perform general maintenance,
- Perform physical measurements,
- Read plans and profiles,
- Recognize unsafe work conditions, and
- Record information.

Class I Sample Questions

1. Pumps typically are categorized as either

   a) Kinetic energy pumps or negative-displacement pumps
   b) Kinetic energy pumps or positive-displacement pumps
   c) Hydraulic energy pumps or positive-displacement pumps
   d) Direct energy pumps or positive-displacement pumps

2. Short-lived transient pressure conditions in force mains and pressure pipes is known as

   a) Water hammer
   b) Normal flow
   c) High velocity
   d) Reverse flow

3. In a sewer lift station, electric motors are designed for either __________ current or __________ current.

   a) Right or left
   b) Positive or negative
   c) Alternating or direct
   d) Direct or indirect

4. What is the "Ten State Standards" minimum ventilation requirement in a dry well of a sewer lift station?

   a) 30 air changes per hour intermittently or six air changes per hour continuously
   b) 20 air changes per hour intermittently or nine air changes per hour continuously
   c) 35 air changes per hour intermittently or 12 air changes per hour continuously
   d) 25 air changes per hour intermittently or 10 air changes per hour continuously

5. What does SCADA stand for?

   a) Supervisory contact and data acquisition
   b) Supervisory control and data alternative
   c) Supervisory center and display availability
   d) Supervisory control and data acquisition

6. Pump capacity is typically expressed in

   a) Gallons per minute (gpm) or million gallons per day (mgd) (liters per minute [L/m] or million liters per day [mil. L/d])

b) Feet per second (ft/sec) or pound-force per square inch (psi) (meter per second [m/sec] or kilogram-force per square centimeter [kgf/cm^2])

c) Feet per second squared (ft/sec^2) or pound-force per square inch (psi) (meter per second squared [m/sec^2] or kilogram-force per square centimeter [kgf/cm^2])

d) Pound-force per square inch (psi) or shaft power (horsepower [hp]) (kilogram-force per square centimeter [kgf/cm^2] or kilowatts [kW])

Answers

1. **Answer:** B

Pumps typically are categorized as kinetic energy pumps or positive-displacement pumps. Kinetic energy pumps produce high-velocity kinetic energy in the pump that is converted to pressure. Positive-displacement pumps produce pressure by applying force directly to the fluid.

Reference: Water Environment Federation (2004) *Design of Wastewater and Stormwater Pumping Stations;* Manual of Practice No. FD-4; Water Environment Federation: Alexandria, Virginia; p 64.

2. **Answer:** A

Rapid changes in flow can lead to pipe pressure upsurges or downsurges that can cause rupture or water column separation and collapse of the pipe. "Water hammer" is the name given to the rapid fall and subsequence rise of pressure in a pipeline that can result from rapid changes in flow.

Reference: Water Environment Federation (2004) *Design of Wastewater and Stormwater Pumping Stations;* Manual of Practice No. FD-4; Water Environment Federation: Alexandria, Virginia; p 134.

3. **Answer:** C

Electric motors are designed for either alternating current or direct current. Direct current motors seldom are used in municipal systems, where they would only be used in variable-speed systems. Alternating current motors can be either single phase or three phases.

Reference: Water Environment Federation (2004) *Design of Wastewater and Stormwater Pumping Stations;* Manual of Practice No. FD-4; Water Environment Federation: Alexandria, Virginia; p 94.

4. **Answer:** A

As per the "Ten State Standards," the minimum ventilation for dry wells of wastewater pumping stations is 30 air changes per hour intermittently or six air changes per hour continuously.

Reference: Water Environment Federation (2004) *Design of Wastewater and Stormwater Pumping Stations;* Manual of Practice No. FD-4; Water Environment Federation: Alexandria, Virginia; p 268.

5. **Answer:** D

At the central facility, the supervisory control and data acquisition system provides interactive displays for status monitoring and control, alarm processing, trending of analog variables such as flow, report generation, and historical data collection.

Reference: Water Environment Federation (2004) *Design of Wastewater and Stormwater Pumping Stations;* Manual of Practice No. FD-4; Water Environment Federation: Alexandria, Virginia; p 226.

6. **Answer:** A

Capacity is the volume of liquid pumped in a unit of time. The capacity typically is expressed as (gallons per minute [gpm]) or million gallons per day (mgd) (liters per minute [L/m] or million liters per day [mil. L/d]).

Reference: Water Environment Federation (2004) *Design of Wastewater and Stormwater Pumping Stations;* Manual of Practice No. FD-4; Water Environment Federation: Alexandria, Virginia; p 61.

Class II Sample Questions

1. What is the range of the optimum velocity in force mains?

   a) 1 and 2 ft/sec (0.3 and 0.6 m/s)
   b) 3 and 5 ft/sec (0.9 and 1.5 m/s)
   c) 6 and 7 ft/sec (1.8 and 2.1 m/s)
   d) 8 and 10 ft/sec (2.4 and 3.0 m/s)

2. A lightning arrester is

   a) A protective device for limiting the high surge voltages on equipment by releasing a surge current to the atmosphere
   b) A protective device for limiting the high surge voltages on equipment by releasing a surge current to the ground
   c) A protective device for limiting the high surge voltages on equipment by releasing a surge current to the electric grid
   d) A protective device for limiting the low surge voltages on equipment by releasing a surge current to the electric grid

3. The term, *static head,* is defined as

   a) The amount of energy produced by a pump at high velocity
   b) The difference in elevation between the bottom of the wet well and the centerline of the pump

c) The positive and negative displacement of the discharge pipe
d) The vertical distance from a specific point of the water surface when the water is not moving

4. When does cavitation in a pump occur?

   a) When the net positive suction head (NPSH) required by the pump at the eye of the impeller is less than the NPSH available in the system
   b) When the negative pressure of the inlet drops below the atmospheric pressure of the fluid being pumped
   c) When the absolute pressure at the eye of the impeller is less than the vapor pressure of the fluid being pumped, causing the fluid to boil and form vapor bubbles
   d) When the absolute pressure at the eye of the impeller is higher than the vapor pressure of the fluid being pumped, causing the fluid to boil and form vapor bubbles

5. Air-release valves should be considered for installation at _____ on discharge piping.

   a) Midpoint
   b) High points
   c) Discharge manhole
   d) Straight level point

Answers

1. **Answer:** B

Optimum velocities for reducing cost and preventing accumulation of solids range between 3 and 5 ft/sec (0.9 and 1.5 m/s), which is high enough to resuspend material that may settle out when the pumps are not operating.

Reference: Water Environment Federation (2004) *Design of Wastewater and Stormwater Pumping Stations;* Manual of Practice No. FD-4; Water Environment Federation: Alexandria, Virginia; p 125.

2. **Answer:** B

A protective device for limiting the high surge voltages on equipment by releasing a surge current to the ground. Grounding of the station will affect system reliability.

Reference: Water Environment Federation (2004) *Design of Wastewater and Stormwater Pumping Stations;* Manual of Practice No. FD-4; Water Environment Federation: Alexandria, Virginia; pp 162, 171, and 172.

3. **Answer:** D

California State University, Sacramento (2003) *Operation and Maintenance of Wastewater Collection Systems*, 6th ed.; California State University: Sacramento, California; Volume I, p 710.

4. **Answer:** C

Cavitation in a pump occurs when the absolute pressure at the eye of the impeller is less than the vapor pressure of the fluid pumped, causing the fluid to boil and form vapor bubbles. Cavitation results in the loss of pump capacity and efficiency and causes physical damage to the pump.

Reference: Water Environment Federation (2004) *Design of Wastewater and Stormwater Pumping Stations;* Manual of Practice No. FD-4; Water Environment Federation: Alexandria, Virginia; p 71.

5. **Answer:** B

Air-release valves should be considered for installation at high points in discharge piping. Air valves will release trapped air when putting the line in service and during regular operation.

Reference: Water Environment Federation (2004) *Design of Wastewater and Stormwater Pumping Stations;* Manual of Practice No. FD-4; Water Environment Federation: Alexandria, Virginia; p 124.

Class III Sample Questions

1. For safety purposes, __________ should always be present during a lift station visit.

   a) A operational work truck
   b) 5 gal (0.02 m^3) of potable water
   c) Two operators
   d) A paramedic

2. A __________ is required at each lift station and should include, at a minimum, the date, time in, time out, and name of the visitor.

   a) Radio communication
   b) Visiting guest sheet
   c) Sign-in log
   d) Work order

3. A lift station book should contain all of the following **EXCEPT**:

   a) Lift station identification number or name
   b) Lift station description
   c) Name and registration number of the design engineer
   d) Responses to emergency conditions

4. Maintenance records are an important part of a lift station's operation and maintenance program and should be filed

   a) At the agency's main office
   b) In your work truck
   c) In the round file
   d) In the control panel cabinet

5. A remote terminal unit (RTU) receives analog and digital

   a) Voice commands
   b) Video displays
   c) Inputs
   d) Web casts

6. A __________ must never be bypassed because it is the only protection a circuit has.

   a) Process step
   b) Supervisor
   c) Vehicle
   d) Fuse

7. Pump head is measured

   a) At the same points and under the same conditions of the pump system
   b) At the same point's only system
   c) Under the same conditions of the pump system only
   d) By the difference in the elevation of a liquid surface above or below a reference datum elevation

Answers

1. **Answer:** C

   Reference: California State University, Sacramento (2010) *Operation and Maintenance of Wastewater Collection Systems*, 7th ed.; California State University: Sacramento, California; Volume II, p 94.

2. **Answer:** C

   Reference: California State University, Sacramento (2010) *Operation and Maintenance of Wastewater Collection Systems*, 7th ed.; California State University: Sacramento, California; Volume II, p 102.

3. **Answer:** C

   Reference: California State University, Sacramento (2010) *Operation and Maintenance of Wastewater Collection Systems*, 7th ed.; California State University: Sacramento, California; Volume II, p 103.

4. **Answer:** A

   Reference: California State University, Sacramento (2010) *Operation and Maintenance of Wastewater Collection Systems*, 7th ed.; California State University: Sacramento, California; Volume II, p 112.

5. **Answer:** C

   Reference: California State University, Sacramento (2010) *Operation and Maintenance of Wastewater Collection Systems*, 7th ed.; California State University: Sacramento, California; Volume II, p 96.

6. **Answer:** D

   Reference: California State University, Sacramento (2010) *Operation and Maintenance of Wastewater Collection Systems*, 7th ed.; California State University: Sacramento, California; Volume II, p 162.

7. **Answer:** D

   Reference: Water Environment Federation (2004) *Design of Wastewater and Stormwater Pumping Stations;* Manual of Practice No. FD-4; Water Environment Federation: Alexandria, Virginia; p 61.

Class IV Sample Questions

1. What are three reasons why a pump will fail to function?

   a) The log book was not correctly filled out, failure in the electrical or control system, and failure in the mechanical system
   b) Debris plugging the impellor, failure in the mechanical system, and the log book was not correctly filled out
   c) Failure of the electrical or control system, debris plugging the impellor, and the log book was not filled out correctly
   d) A failure of the electrical or control system, failure in the mechanical system, and debris plugging the impellor

2. During the first year of operation of new equipment, the __________ recommendations must be closely followed to maintain __________.

   a) Equipment supplier's/warranty
   b) Millwright's/optimum performance
   c) Utility management's/budget requirements
   d) Manufacturer's/warranty

3. A __________ can change the speed of a pump motor to allow for the pump to match flowrates into the wet well.

   a) Fused disconnect
   b) Hand-off-auto (HOA) switch
   c) Variable-frequency drive (VFD)
   d) Drive shoe brake (DSB)

4. You are making an inspection of a pumping station that includes a below-grade wet well. The wet well cover is a hinged steel cover at grade. Before you open the cover to make a visual observation, what steps must you take?

   a) Secure the perimeter area of the wet well with an appropriate area barrier if one does not already exist, securing yourself with a fall protection device.
   b) Open the cover and kneel down so that you can put your head in as far as possible in order to see as clearly as possible.
   c) Shut off all power to the pumping station so no one gets electrocuted.
   d) Never open the cover to a below-grade wet well.

5. Which three pumps are used to pump wastewater?

   a) Incline screw, horizontal no-clog, submersible
   b) Submersible, horizontal no-clog, low vortex reliance
   c) Horizontal no-clog, low vortex reliance, incline screw
   d) Low vortex reliance, submersible, incline screw

Answers

1. **Answer:** D

   Reference: California State University, Sacramento (2010) *Operation and Maintenance of Wastewater Collection Systems*, 7th ed.; California State University: Sacramento, California; Volume II, p 94.

2. **Answer:** D

   Reference: California State University, Sacramento (2010) *Operation and Maintenance of Wastewater Collection Systems*, 7th ed.; California State University: Sacramento, California; Volume II, p 109.

3. **Answer:** C

   Reference: California State University, Sacramento (2010) *Operation and Maintenance of Wastewater Collection Systems*, 7th ed.; California State University: Sacramento, California; Volume II, p 35.

4. Answer: A

Reference: California State University, Sacramento (2003) *Operation and Maintenance of Wastewater Collection Systems*, 6th ed.; California State University: Sacramento, California; Volume I, p 139.

5. Answer: A

Reference: California State University, Sacramento (2010) *Operation and Maintenance of Wastewater Collection Systems*, 7th ed.; California State University: Sacramento, California; Volume II, p 199.

Monitor, Evaluate, and Adjust Collection System

	Class I	Class II	Class III	Class IV
Aeration for hydrogen sulfide control	Comprehension	Comprehension	Application	Application
Biological filters for odor control	N/A	Comprehension	Application	Application
Chemical addition for hydrogen sulfide control	Comprehension	Comprehension	Application	Analysis
Cross connections	Comprehension	Comprehension	Comprehension	Comprehension
Flow monitoring	Comprehension	Comprehension	Analysis	Analysis
Force mains	Analysis	Analysis	Analysis	Analysis
Gravity sewers	Analysis	Analysis	Analysis	Analysis
Infiltration (inflow, exfiltration)	Application	Application	Analysis	Analysis
Lift stations	Analysis	Analysis	Analysis	Analysis
Manholes	Analysis	Analysis	Analysis	Analysis
Measuring and control systems	Analysis	Analysis	Analysis	Analysis
Pressure sewers (septic tank effluent pump, or STEP, systems)	N/A	N/A	N/A	Analysis
Vacuum sewers	N/A	Analysis	Analysis	Analysis

Required Capabilities

Knowledge of

- Adjusting equipment,
- Assessing likelihood of disaster occurring,
- Calibrating equipment,
- Communicating in writing,
- Communicating verbally,
- Conducting meetings,
- Conducting training programs,
- Coordinating emergency response with other organizations,
- Demonstrating safe work habits,
- Diagnosing/troubleshooting,
- Differentiating between preventive and corrective maintenance,
- Discriminating between normal and abnormal conditions,

- Evaluating data from inspections,
- Evaluating operation of equipment,
- Evaluating proposals, and
- Evaluating system performance.

Ability to

- Adjust equipment,
- Assess likelihood of disaster occurring,
- Calibrate equipment,
- Communicate in writing,
- Communicate verbally,
- Conduct meetings,
- Conduct training programs,
- Coordinate emergency response with other organizations,
- Demonstrate safe work habits,
- Diagnose/troubleshoot,
- Differentiate between preventive and corrective maintenance,
- Discriminate between normal and abnormal conditions,
- Evaluate cause of damage,
- Evaluate data from inspections,
- Evaluate operation of equipment,
- Evaluate system performance,
- Evaluate proposals,
- Follow written procedures,
- Identify cause of damage,
- Identify potential safety hazards,
- Interpret data,
- Interpret material safety data sheets,
- Locate cause of flow problems,
- Maintain system in normal operating condition,
- Monitor electrical equipment, and
- Monitor mechanical equipment.

Class I Sample Questions

1. Hydrogen sulfide is generated in collection systems when

a) Chlorine is added after long stretches of gravity sewers
b) Iron salts are mixed with the sulfides in wastewater
c) Oxygen is not present in the wastewater flowing through the system
d) Hydrogen peroxide is added to a pumping station wet well

2. Wastewater is conveyed from a building to a low-pressure small-diameter force main sewer system using a __________.

   a) Screw pump system
   b) Vacuum pump system
   c) Grinder pump system
   d) Air lift pump system

Answers

1. **Answer:** C

   Reference: Water Environment Federation (2008) *Operation of Municipal Wastewater Treatment Plants,* 6th ed.; Manual of Practice No. 11; Water Environment Federation: Alexandria, Virginia; pp 13–15.

2. **Answer:** C

   Reference: California State University, Sacramento (2012) *Small Wastewater Systems Operation and Maintenance*, 2nd ed.; California State University: Sacramento, California; Volume I.

Class II Sample Questions

1. Hydrogen sulfide can be removed by passing wet well exhaust through a vessel containing wood chips mixed with organic matter such as compost. This odor control unit is known as a

   a) Chemical scrubber
   b) Biofilter
   c) Carbon filter
   d) Anthracite filter

2. Measuring devices such as weirs, flumes, or area velocity meters are often permanently or temporarily installed in gravity flow pipes to measure __________.

   a) Nutrient concentrations
   b) Flow
   c) Pressure
   d) Energy use

3. A lift station has a wet well and a dry well. On/off float switches are used to activate fixed speed pump(s). During the pumping cycle, just before the pump turns off, a

noise occurs in the centrifugal pump that sounds like marbles or gravel trapped in the volute. This happens each cycle. What is the likely cause?

a) Cavitation
b) Excess torque
c) Packing too tight
d) Excess fluid flow

4. A primary advantage of a vacuum sewer system is that

a) There are few moving parts in contact with the wastewater
b) There is no electrical consumption
c) The life cycle costs are low
d) The water use required for flushing is low

Answers

1. **Answer:** B

Reference: Water Environment Federation (2008) *Operation of Municipal Wastewater Treatment Plants,* 6th ed.; Manual of Practice No. 11; Water Environment Federation: Alexandria, Virginia; pp 13–36.

2. **Answer:** B

Reference: American Society of Civil Engineers/Environmental and Water Resources Institute; Water Environment Federation (2007) *Gravity Sanitary Sewer Design and Construction,* 2nd ed.; ASCE Manuals and Reports on Engineering Practice No. 60; WEF Manual of Practice No. FD-5; American Society of Civil Engineers: Reston, Virginia; p 50.

3. **Answer:** A

Reference: California State University, Sacramento (2010) *Operation and Maintenance of Wastewater Collection Systems*, 7th ed.; California State University: Sacramento, California; Volume II, p 68.

4. **Answer:** D

Reference: California State University, Sacramento (2012) *Small Wastewater Systems Operation and Maintenance*, 2nd ed.; California State University: Sacramento, California; Volume I.

Class III Sample Questions

1. You have a classic wet well, dry well housing centrifugal pumps, and a force main that terminates at the bar racks at a wastewater plant. Your pump elevation to free water discharge at the bar racks is 55 ft (17 m). Friction head losses are 45 ft (14 m). You have

a double mechanical seal with flushing water. The seal water pressure must exceed the pressure of the pumped wastewater. What seal water pressure is necessary at the seal?

a) 15 psi (103 kPa)
b) 25 psi (172 kPa)
c) 40 psi (276 kPa)
d) 55 psi (379 kPa)

2. You have 2 miles (3.2 km) of gravity sewer that runs parallel to a stream and you suspect infiltration in the spring and exfiltration in the summer. The geology of the area indicates that gravel and sand lenses are common. You have inspected all the manholes, benches, and barrels and there is no apparent infiltration/inflow from these access points when the stream is high. What is an effective way to confirm your suspicion of infiltration or exfiltration?

a) Run a mandrel through the pipe in this section during dry season
b) Dye test this section by introducing dye through a lateral cleanout
c) Closed circuit television of this section during high stream conditions
d) Smoke test this section during spring high stream conditions

3. When compared to a larger-diameter gravity system, what is a key advantage of a collection system in which each customer has a septic tank effluent pump (STEP) system feeding a small-diameter force main to the treatment system?

a) Uses less electricity
b) Little infiltration/inflow
c) Lower life cycle cost
d) Larger flows are treated

Answers

1. Answer: D

Reference: California State University, Sacramento (2010) *Operation and Maintenance of Wastewater Collection Systems*, 7th ed.; California State University: Sacramento, California; Volume II, p 219.

2. Answer: C

Reference: California State University, Sacramento (2010) *Operation and Maintenance of Wastewater Collection Systems*, 7th ed.; California State University: Sacramento, California; Volume II, p 251.

3. Answer: B

Reference: California State University, Sacramento (2012) *Small Wastewater Systems Operation and Maintenance*, 2nd ed.; California State University: Sacramento, California; Volume I.

Class IV Sample Questions

1. In a warm weather climate, a community has a long, flat collector sewer where there is a buildup of dissolved sulfides in the wastewater. The pH is 8.5 and there is no apparent hydrogen sulfide odor. Homes are near the pumping station wet well that this sewer discharges into. Your supervisor wants you to explore methods to reduce potential odors. What three methods can be used?

   a) Installation of a wet chemical scrubber to the wet well vent, addition of ferric chloride to an upstream point in the collection system, and addition of sodium hypochlorite to an upstream point in the collection system
   b) Installation of a wet chemical scrubber to the wet well vent, addition of an acid to an upstream point in the collection system, addition of ferric chloride to an upstream point in the collection system, and addition of sodium hypochlorite to an upstream point in the collection system
   c) Addition of ferric chloride to an upstream point in the collection system, addition of sodium hypochlorite to an upstream point in the collection system, and addition of an acid to an upstream point in the collection system
   d) Addition of sodium hypochlorite to an upstream point in the collection system, addition of an acid to an upstream point in the collection system, installation of a wet chemical scrubber to the wet well vent, and addition of ferric chloride to an upstream point in the collection system

2. A(n) __________ can occur when a sewer blockage occurs in the collection system in a low area where homes are located.

   a) Shift of invert elevation
   b) Increase in water pressure
   c) Decrease in water pressure
   d) Cross connection

Answers

1. **Answer:** A

   Reference: California State University, Sacramento (2010) *Operation and Maintenance of Wastewater Collection Systems*, 7th ed.; California State University: Sacramento, California; Volume II, p 108.

2. **Answer:** D

   Reference: California State University, Sacramento (2003) *Operation and Maintenance of Wastewater Collection Systems*, 6th ed.; California State University: Sacramento, California; Volume I, p 6.

Perform Security, Safety, and Administrative Procedures

	Class I	Class II	Class III	Class IV
Administer system				
Administer safety compliance program	Application	Application	Analysis	Analysis
Develop budget	N/A	N/A	Analysis	Analysis
Develop capital improvement plan	N/A	N/A	N/A	Analysis
Develop operation and maintenance plan	Application	Application	Analysis	Analysis
Evaluate employee performance	Application	Application	Application	Analysis
Hire employees	N/A	N/A	Application	Analysis
Maintain records	Application	Application	Analysis	Analysis
Perform workplace safety evaluations	Application	Application	Analysis	Analysis
Plan and organize work activities	Application	Application	Analysis	Analysis
Record and evaluate data	Application	Application	Analysis	Analysis
Respond to public complaints	Application	Application	Analysis	Analysis
Supervise employee work activities	Application	Application	Application	Analysis
Write reports (federal, internal, state)	Application	Application	Analysis	Analysis
Safety procedures				
Calibration of atmospheric testing devices	Analysis	Analysis	Analysis	Analysis
Chemical spill	Analysis	Analysis	Analysis	Analysis
Confined space entry	Analysis	Analysis	Analysis	Analysis
Electrical hazards	Analysis	Analysis	Analysis	Analysis
Fires	Analysis	Analysis	Analysis	Analysis
First aid	Analysis	Analysis	Analysis	Analysis
Hazardous material	Analysis	Analysis	Analysis	Analysis
Infectious disease	Analysis	Analysis	Analysis	Analysis
Lifting	Analysis	Analysis	Analysis	Analysis
Lockout/tagout	Analysis	Analysis	Analysis	Analysis
Personal protection equipment	Analysis	Analysis	Analysis	Analysis
Respiratory protection	Analysis	Analysis	Analysis	Analysis
Shoring	Analysis	Analysis	Analysis	Analysis
Traffic control	Analysis	Analysis	Analysis	Analysis
Trenching and excavation	Analysis	Analysis	Analysis	Analysis
Emergency plans				
Combined sewer overflows	Application	Application	Analysis	Analysis
Disasters	Application	Application	Analysis	Analysis
Manhole hazards	Application	Application	Analysis	Analysis
Sanitary sewer overflow	Application	Application	Analysis	Analysis
System failure	Application	Application	Analysis	Analysis

Required Capabilities

Knowledge of

- Emergency plans,
- Function of recordkeeping system,
- General electrical principles,
- General hydraulic principles,
- Hazardous situations,
- Hydrogen sulfide generation,
- Methane generation,
- Monitoring and reporting requirements,
- Pipe fittings and joining methods,
- Pipeline construction principles,
- Piping material type and size,
- Pneumatics,
- Potential causes of disasters in the system,
- Potential effect of disasters on the system,
- Regulations,
- Risk management,
- Safety regulations,
- Startup and shut-down procedures,
- System operation and maintenance,
- Types of lift stations,
- Types of pumps,
- Wastewater collection design parameters,
- Wastewater collection system concepts,
- Wastewater collection system operation and maintenance, and
- Wastewater treatment concepts.

Ability to

- Adjust equipment,
- Assess the likelihood of a disaster occurring,
- Calibrate equipment,
- Communicate in writing,
- Communicate verbally,
- Conduct meetings,
- Conduct training programs,
- Coordinate emergency response with other organizations,
- Demonstrate safe work habits,
- Diagnose/troubleshoot,

- Differentiate between preventive and corrective maintenance,
- Discriminate between normal and abnormal conditions,
- Evaluate cause of damage,
- Evaluate data from inspections,
- Evaluate equipment operation,
- Evaluate proposals,
- Evaluate system performance,
- Follow written procedures.
- Generate written safety procedures,
- Identify cause of damage,
- Identify potential safety hazards,
- Interpret data,
- Interpret material safety data sheets,
- Locate cause of flow problems,
- Maintain system in normal operating condition,
- Monitor mechanical equipment,
- Negotiate contracts,
- Operate safety equipment,
- Order necessary spare parts,
- Perform general maintenance,
- Perform impact assessments,
- Perform mathematical calculations,
- Perform physical measurements,
- Prepare proposals,
- Read plans and profiles,
- Recognize unsafe work conditions,
- Record information,
- Review reports,
- Select safety equipment,
- Transcribe data,
- Translate technical language into common terminology, and
- Write policies and procedures.

Class I Sample Questions

1. What is the most important concern after a discharge of hazardous material into a wastewater collection system?

a) Finding out who caused the discharge
b) Determining where to buy the proper cleanup equipment
c) Minimizing health risks
d) Preventing odors

2. When a complaint call is received from a customer, the steps to efficiently resolve the issue would be to first prepare a thorough record of the call and then

   a) File the report for future reference.
   b) Respond to the complaint as soon as possible.
   c) Check with the city manager before responding.
   d) Refer the customer to a contract/professional who can help.

3. You are in charge of a large operation with four crew leaders; the work of three of these crew leaders is exceptionally good, but that of the fourth is substandard. What should you do?

   a) Demote the substandard crew leader and bring up a replacement from the ranks.
   b) Discuss the problem with the substandard crew leader and offer to help.
   c) Find a replacement and then fire the substandard crew leader.
   d) Wait to see if the substandard crew leader does better.

4. Which of the following questions is unacceptable to ask during a pre-employment interview?

   a) Can you complete the tasks required at this position?
   b) How long have you been a resident of this state?
   c) Are you a United States citizen?
   d) What is your past work experience?

5. Rotating operators to different crews or types of work should be done to

   a) Determine how easy it is for new operators to operate equipment
   b) Promote operators that are capable to get work done
   c) Improve efficiency in the department
   d) Prevent operators from being promoted

Answers

1. **Answer:** C

   Reference: Water Environment Federation (2012) *Safety, Health, and Security in Wastewater Systems*, 6th ed.; Manual of Practice No. 1; Water Environment Federation: Alexandria, Virginia; pp 188 and 279.

2. **Answer:** B

   Reference: California State University, Sacramento (2010) *Operation and Maintenance of Wastewater Collection Systems*, 7th ed.; California State University: Sacramento, California; Volume II, Section 12.9, p 519.

3. **Answer:** B

Reference: California State University, Sacramento (2010) *Operation and Maintenance of Wastewater Collection Systems*, 7th ed.; California State University: Sacramento, California; Volume II, Section 14.24, pp 613–614.

4. **Answer:** B

Reference: California State University, Sacramento (2010) *Operation and Maintenance of Wastewater Collection Systems*, 7th ed.; California State University: Sacramento, California; Volume II, pp 436–438.

5. **Answer:** C

Reference: California State University, Sacramento (2010) *Operation and Maintenance of Wastewater Collection Systems*, 7th ed.; California State University: Sacramento, California; Volume II, pp 436–438.

Class II Sample Questions

1. The individual responsible for signing the confined space entry permit before entry and canceling it following entry is known as the

   a) Attendant
   b) Entry supervisor
   c) Safety officer
   d) Superintendent

2. An oxygen alarm reading of less than 19.5% on a gas detector during confined space entry is known as a(n) __________ reading.

   a) Oxygen-enrichment
   b) Lower explosion limit
   c) Oxygen-deficiency
   d) Hazardous substance

3. When identifying a confined space, the Occupational Safety and Health Administration (OSHA) standard requires the employer to

   a) Limit access and egress
   b) Eliminate dark and damp areas
   c) Identify employee job duties
   d) Establish a verbal procedure for issuance, use, and cancellation of entry permits

4. The numerical designation for neutrality when measuring pH is

a) 0
b) 2
c) 7
d) 10

5. Who is the only person who should remove the lockout/tagout devices when finished working on electrical systems?

a) The supervisor
b) The person who assigned the work
c) The person who installed the devices
d) The helper

6. When sloping of an excavation made in Class B soil is not possible because of limited work space, what is the minimum depth at which shoring is always required?

a) 3 ft (0.9 m)
b) 4 ft (1.2 m)
c) 5 ft (1.5 m)
d) 7 ft (2.1 m)

7. The term given to the person responsible for inspection and evaluation of the safety of a trench or excavation is

a) Supervisor
b) Competent person
c) Excavator
d) Superintendent

8. In a 7-ft (2.1-m) deep trench, what is the maximum distance allowed for a worker to reach a means of egress?

a) 10 ft (3 m)
b) 15 ft (4.6 m)
c) 25 ft (7.6 m)
d) 30 ft (9.1 m)

Answers

1. **Answer:** B

   Reference: Water Environment Federation (2012) *Safety, Health, and Security in Wastewater Systems*, 6th ed.; Manual of Practice No. 1; Water Environment Federation: Alexandria, Virginia; pp 174–176.

   Additional reference: California State University, Sacramento (2003) *Operation and Maintenance of Wastewater Collection Systems*, 7th ed.; California State University: Sacramento, California; Volume II, Chapter 11, Section 11.814, p 361.

2. **Answer:** C

   Reference: Water Environment Federation (2012) *Safety, Health, and Security in Wastewater Systems*, 6th ed.; Manual of Practice No. 1; Water Environment Federation: Alexandria, Virginia; p 415.

   Additional reference: California State University, Sacramento (2003) *Operation and Maintenance of Wastewater Collection Systems*, 7th ed.; California State University: Sacramento, California; Volume II, Chapter 11, Section 11.1, p 346.

3. **Answer:** C

   Reference: Water Environment Federation (2012) *Safety, Health, and Security in Wastewater Systems*, 6th ed.; Manual of Practice No. 1; Water Environment Federation: Alexandria, Virginia; pp 172 and 185.

4. **Answer:** C

   Reference: Water Environment Federation (2012) *Safety, Health, and Security in Wastewater Systems*, 6th ed.; Manual of Practice No. 1; Water Environment Federation: Alexandria, Virginia; p 416.

5. **Answer:** C

   Reference: Water Environment Federation (2012) *Safety, Health, and Security in Wastewater Systems*, 6th ed.; Manual of Practice No. 1; Water Environment Federation: Alexandria, Virginia; pp 145 and 203.

   Additional reference: California State University, Sacramento (2003) *Operation and Maintenance of Wastewater Collection Systems*, 6th ed.; California State University: Sacramento, California; Volume I, Chapter 4, Section 4.9, p 159.

6. **Answer:** C

   Reference: Occupational Safety and Health Administration (1995) 29 CFR PART 1926, Subpart P, 1926.652 a1ii; Occupational Safety and Health Administration: Washington, D.C.; p 263.

Additional reference: California State University, Sacramento (2003) *Operation and Maintenance of Wastewater Collection Systems*, 6th ed.; California State University: Sacramento, California; Volume I, Chapter 4, Section 4.8, p 149.

7. **Answer:** B

"Competent person" means one who is capable of identifying existing and predictable hazards in the surroundings, or working conditions which are unsanitary, hazardous, or dangerous to employees, and who has authorization to take prompt corrective measures to eliminate them.

Reference: Occupational Safety and Health Administration (1995) 29 CFR PART 1926, Subpart P, 1926.650 b; Occupational Safety and Health Administration: Washington, D.C.; p 261.

8. **Answer:** C

Reference: Occupational Safety and Health Administration (1995) 29 CFR PART 1926, Subpart P, 1926.651 c2; Occupational Safety and Health Administration: Washington, D.C.; p 262.

Class III Sample Questions

1. The Workers' Right-to-Know laws address

   a) Hazard communication
   b) Personnel policies
   c) Union contract issues
   d) Trenching and shoring safety

2. The five important areas of traffic control are: advance warning area, transition area, buffer space, __________, and termination area.

   a) Confined space
   b) Work area
   c) Parking area
   d) Caution area

3. One of the primary safety concerns of working in combined sewer systems is

   a) Extreme changes in flowrates caused by storm runoff
   b) The presence of debris from storm inlets
   c) The size of the pipes
   d) The material of the pipe

4. What is the main reason to check an applicant's references?

   a) To verify past work experiences

- **b)** To verify a criminal record
- **c)** To verify educational background
- **d)** To verify an on-the-job injury claim

5. A good public relations program consists of professional appearance, courteous customer service, and __________.

- **a)** Providing equal service to all customers
- **b)** Having a professional public relations person on staff
- **c)** Promoting cost savings by keeping obsolete equipment in service
- **d)** Withholding cost of backups from the public

Answers

1. Answer: A

Reference: California State University, Sacramento (2003) *Operation and Maintenance of Wastewater Collection Systems*, 6th ed.; California State University: Sacramento, California; Volume I, Chapter 4, Section 4.12, p 165.

2. Answer: B

Reference: California State University, Sacramento (2003) *Operation and Maintenance of Wastewater Collection Systems*, 6th ed.; California State University: Sacramento, California; Volume I, Chapter 4, Section 4.3, p 95.

3. Answer: A

Reference: Water Environment Federation (2012) *Safety, Health, and Security in Wastewater Systems*, 6th ed.; Manual of Practice No. 1; Water Environment Federation: Alexandria, Virginia; p 217.

Additional reference: California State University, Sacramento (2003) *Operation and Maintenance of Wastewater Collection Systems*, 6th ed.; California State University: Sacramento, California; Volume I, Chapter 14, Section 14.00, p 605.

4. Answer: A

Reference: California State University, Sacramento (2010) *Operation and Maintenance of Wastewater Collection Systems*, 7th ed.; California State University: Sacramento, California; Volume II, Chapter 12, Section 12312, pp 438–439.

5. Answer: A

Reference: California State University, Sacramento (2010) *Operation and Maintenance of Wastewater Collection Systems*, 7th ed.; California State University: Sacramento, California; Volume II, p 520.

Class IV Sample Questions

1. Occupational Safety and Health Administration (OSHA) standards require that medical services be available "in near proximity". "In near proximity" has been defined as

   a) 1 to 2 minutes
   b) 3 to 4 minutes
   c) 5 to 10 minutes
   d) 10 to 15 minutes

2. The two agencies that typically approve personal protection equipment are the American National Standards Institute (ANSI) and

   a) Occupational Safety and Health Administration (OSHA)
   b) U.S. Environmental Protection Agency (U.S. EPA)
   c) National Institute for Occupational Safety and Health (NIOSH)
   d) American Society for Testing and Materials (ASTM)

3. The intrastate initiative to provide mutual aid to wastewater agencies during a disaster is known as

   a) OSHA
   b) Water and Wastewater Agency Response Network (WARN)
   c) U.S. EPA
   d) NIOSH

4. Collection system managers keep good records to

   a) Stay in compliance with regulations
   b) Justify the need for a work order management program
   c) Maintain the level of work for staff
   d) Help solve similar problems the next time

5. When can work reports be disposed of?

   a) When the manager retires
   b) When reports have been entered into a recordkeeping database
   c) When you have run out of money
   d) As soon as the work has been completed

6. Which of the following situations would be the highest priority?

   a) Home service line backed up
   b) Mainline flooded basement of a home
   c) Cleanout is overflowing

d) Manhole is overflowing

Answers

1. **Answer:** B

Reference: Water Environment Federation (2010) *Wastewater Collection Systems Management*, 6th ed.; Manual of Practice No. 7; Water Environment Federation: Alexandria, Virginia; p 213.

2. **Answer:** C

Reference: Water Environment Federation (2010) *Wastewater Collection Systems Management*, 6th ed.; Manual of Practice No. 7; Water Environment Federation: Alexandria, Virginia; p 213.

3. **Answer:** B

Reference: Water Environment Federation (2010) *Wastewater Collection Systems Management*, 6th ed.; Manual of Practice No. 7; Water Environment Federation: Alexandria, Virginia; p 382.

4. **Answer:** D

Reference: California State University, Sacramento (2010) *Operation and Maintenance of Wastewater Collection Systems*, 7th ed.; California State University: Sacramento, California; Volume II, p 650.

5. **Answer:** B

Reference: California State University, Sacramento (2010) *Operation and Maintenance of Wastewater Collection Systems*, 7th ed.; California State University: Sacramento, California; Volume II, p 471.

6. **Answer:** A

Reference: California State University, Sacramento (2010) *Operation and Maintenance of Wastewater Collection Systems*, 7th ed.; California State University: Sacramento, California; Volume II, pp 428 and 429.

www.ingramcontent.com/pod-product-compliance
Lightning Source LLC
Chambersburg PA
CBHW081043200925
32899CB00006B/65
* 9 7 8 1 5 7 2 7 8 3 0 1 0 *